Design Make Play for Equity, Inclusion, and Agency

This pioneering book offers a resource for educators, policymakers, researchers, exhibit designers, and program developers that illuminates creative, cutting-edge ways to inspire, engage, and motivate young people about STEM learning in both informal and formal education settings.

A follow-up to the popular book *Design, Make, Play* (2013), this volume combines new research, innovative case studies, and practical advice from the New York Hall of Science (NYSCI) to define and illustrate a vision for creative and immersive learning, focusing on STEM learning experiences that are truly equitable and inclusive, and that foster learners' agency.

Featuring contributions from program developers, facilitators, educators, exhibit designers, and researchers, the book provides real-world examples from informal and formal settings that fill the need for high-quality STEM learning opportunities that are accessible to all learners, including groups underrepresented in STEM education and careers. Chapters of the book describe strategies such as using narratives to make engineering learning more inclusive, engaging English language learners in digital design, focusing on whole-family learning, and introducing underserved students to computational thinking through an immersive computer game.

This book offers both a challenge and a guide to all STEM educators in museums, science centers, and other informal and formal education settings who are seeking out ambitious and more equitable forms of engagement. With leading-edge research and practical advice, the book provides appealing and accessible forms of engagement that will support a diverse range of audiences and deepen their approach to creative STEM learning.

Harouna Ba is Director of the Sara Lee Schupf Family Center for Play, Science, and Technology Learning at the New York Hall of Science (NYSCI), as well as a senior research scientist specializing in children's STEM learning across formal and informal educational settings.

Katherine McMillan Culp is Chief Learning Officer at the New York Hall of Science (NYSCI), where she oversees research, exhibition, and program development; educational and community outreach; and youth development programs.

Margaret Honey is President and CEO of the New York Hall of Science (NYSCI) and leads the museum's commitment to nurturing a generation of creative and collaborative problem solvers in STEM.

Design Make Play for Equity, Inclusion, and Agency

The Evolving Landscape of Creative STEM Learning

Edited by
Harouna Ba, Katherine McMillan Culp, and
Margaret Honey

Routledge
Taylor & Francis Group

NEW YORK AND LONDON

First published 2022
by Routledge
605 Third Avenue, New York, NY 10158

and by Routledge
2 Park Square, Milton Park, Abingdon, Oxon OX14 4RN

Routledge is an imprint of the Taylor & Francis Group, an informa business

Library of Congress Cataloging-in-Publication Data
Names: Ba, Harouna, editor. | Culp, Katherine McMillan, editor. |
Honey, Margaret, editor.
Title: Design make play for equity, inclusion, and agency:
The evolving landscape of creative STEM learning /
edited by Harouna Ba, Katherine McMillan Culp and Margaret Honey.
Description: New York, NY: Routledge, 2021. |
Includes bibliographical references and index.
Identifiers: LCCN 2021003929 | ISBN 9781138572119 (hardback) |
ISBN 9781138572126 (paperback) | ISBN 9780203702345 (ebook)
Doi: 10.4324/9780203702345
Subjects: LCSH: Science–Study and teaching–Activity programs. |
Technology–Study and teaching–Activity programs. |
Engineering–Study and teaching–Activity programs. |
Mathematics–Study and teaching–Activity programs. | Inclusive education.
Classification: LCC Q181 .D3689 2021 | DDC 507.1–dc23
LC record available at https://lccn.loc.gov/2021003929

ISBN: 978-1-138-57211-9 (hbk)
ISBN: 978-1-138-57212-6 (pbk)
ISBN: 978-0-203-70234-5 (ebk)

Typeset in ScalaSans
by Newgen Publishing UK

Contents

List of Figures   viii
List of Tables   xi
About the Contributors   xii
Acknowledgments   xx

Introduction: Design Make Play for Equity, Inclusion,
and Agency   1

HAROUNA BA, KATHERINE MCMILLAN CULP, AND MARGARET HONEY

Part I   Designing for Visitors' Agency   11

1   Designing for Agency in Informal STEM Learning
    Environments   13

    SUSAN M. LETOURNEAU

2   From Explaining to Engaging Visitors: Transforming the
    Facilitator's Role   28

    PRIYA MOHABIR, DOROTHY BENNETT, C. JAMES LIU, AND DAISY TETECATL

3 Narratives, Empathy, and Engineering: Creating Inclusive
   Engineering Activities   44

   SUSAN M. LETOURNEAU, DOROTHY BENNETT, AMELIA MERKER,
   SATBIR MULTANI, C. JAMES LIU, YESSENIA ARGUDO, AND DANA SCHLOSS

4 Co-Designing Learning Dashboards for Informal
   Educators   61

   ELHAM BEHESHTI, LEILAH LYONS, ADITI MALLAVARAPU,
   WREN THOMPSON, BETTY WALLINGFORD, AND STEPHEN UZZO

Part II   Relinquishing Power and Authority in Informal
          Settings   81

5 Museum–Community Engagement to Support
   STEM Learning   83

   ANDRÉS HENRÍQUEZ AND MARCIA BUENO

6 Big Data for Little Kids: Developing an Inclusive
   Program for Young Learners and Their Families   100

   C. JAMES LIU, KATE MASCHAK, DELIA MEZA, SUSAN M. LETOURNEAU,
   AND YESSENIA ARGUDO

7 Designing Maker Programs for Family Engagement   120

   DAVID WELLS, SUSAN M. LETOURNEAU, AND SAMANTHA TUMOLO

8 Innovation Institute: Follow the Youth   134

   DAVID WELLS, ELHAM BEHESHTI, AND DANNY KIRK

Part III   Playing and Learning Across Settings   153

9 See, Touch, and Feel Math: Digital Design for
   English Language Learners   155

   DOROTHY BENNETT, TARA CHUDOBA, XIOMARA FLOWERS,
   AND HEIDI SLOUFFMAN

10 Learning Physics through Embodied Play in a
    School Setting   173

    HAROUNA BA, CHRISTINA O'MALLEY, YESSENIA ARGUDO,
    AND LAYCCA UMER

11 Integrating Computational Thinking Across the
    Elementary Curriculum: A Professional Development
    Approach   195

    ANTHONY NEGRON

12 The Pack: Playfully Embodying Computational
    and Systems Thinking   208

    LEILAH LYONS, STEPHEN UZZO, HAROUNA BA, AND WREN THOMPSON

    Index   226

Figures

1.1 Children collectively write down their individual measurements
on a shared poster-size image of the exhibit Celestial Mechanics   15
1.2 A caregiver helps her child make sense of the collectively
gathered measurements of exhibits to decide which exhibit is
the smallest or biggest   22
2.1 List of habits of mind distributed to youth employed at the museum   34
2.2 Talk-back boards in the Explainer Lounge where Explainers
reflect on how they put habits of mind into action on the
museum floor   36
2.3 Explainer following a visitor's lead in our Design Lab's dowels
structure activity   38
3.1 Testing out a dowel structure built to withstand an earthquake   47
3.2 A child's creation in Help Grandma: a "robot glasses fetcher" to
seek out, grab, and retrieve Nonna's glasses for her   54
3.3 Testing stations in an air-powered vehicles activity, reimagined to use
realistic textured terrains including a tundra (left) and a forest (right) 54
3.4 A narrative activity called Help the Pets   55
4.1 The Connected Worlds exhibit   64
4.2 Examples of participatory design session materials and activities   69
4.3 Task labels created by Explainers in the first participatory design
session that express the desire to see and show all plants and
animals in one biome   70

4.4	Materials for participatory design session 3, including cards representing available types of data and data transformations and printed examples of common types of data visualizations	72
4.5	Initial rough sketches of a Pokédex-inspired interface (top) and a clearer illustration (bottom). Users could explore which pairs of plants and animals occur simultaneously in the same biome to determine whether a particular plant plays a role in forming an animal's habitat	73
5.1	A parent and her child engaged in a design activity	91
5.2	High school STEM fair for community youth	93
6.1	Children showcasing their exhibit prototypes with museum visitors and staff	105
6.2	Percentage of caregiver–child talks devoted to particular data modeling topics	108
6.3	Percentage of caregiver–child talks demonstrating caregiver interaction styles	108
6.4	Data recorded by a child (this page) and her caregiver (next page) during the Measuring James activity	114
6.5	A child and his caregiver working together to measure an exhibit	116
7.1	Children learn how to use a scroll saw in Maker Space	124
8.1	Our theoretical model is situated at the intersection of three core elements: computational making, design thinking, and social entrepreneurship	137
8.2	The program flow	138
8.3	Collaborative relationships to build agency among researchers, facilitators, and participants	140
9.1	Mariana's translation symmetry animation. Each of the faces was plotted based on coordinates she had written down to be precisely symmetrical	162
9.2	Mariana's Ferris wheel project. The top line graph shows the angle of rotation for her Ferris wheel decreasing at each time point, making it slowly rotate as each character boards the wheel	163
9.3	A dog and an owl merged together using the Fraction Mash app	164
9.4	Alejandro's multiple-jointed symmetrical character designed in the Choreo Graph app	165
10.1	NYSCI's Science Playground	177
10.2	App and users	178
11.1	PS 13Q fifth-grade students working on The Case of the Missing Winner activity under the supervision of NYSCI's manager of digital programming	203
12.1	Screenshot from the second prototype, depicting the player's avatar standing in front of an algorithm	209
12.2	Depictions of the six different elemental characters	216

12.3   Screenshots of the first prototype, showing how the environment
       changes as different numbers of different elemental creatures are
       added to it   217
12.4   The algorithm design screen, where players can access stored
       algorithms, amend them, rename them, and create new ones   219

Tables

3.1	Indicators of empathy in observational and interview data	50
3.2	Engineering practices coded for in observational and interview data	50
4.1	Structure of participatory design (PD) sessions	67
6.1	Activities and guiding themes of workshops	104
6.2	Coding scheme for data modeling concepts in parent–child conversation	107
8.1	Program modules for the boot camp and first semester	145
11.1	Computational thinking questioning strategies	201

About the Contributors

Yessenia Argudo is a research and development assistant at the New York Hall of Science (NYSCI). She started her career at NYSCI right after graduating with an MPH in community health education from the City University of New York (CUNY) School of Public Health and Public Policy. She has been involved in a variety of research projects throughout her time with the museum and believes that research is easily transferable throughout fields and topics with the proper knowledge and skills. Yessenia aspires to become a well-rounded professional and develop as an educator and researcher by gaining experience with as many different methodologies as possible.

Harouna Ba is a senior research scientist and director of the Sara Lee Schupf Family Center for Play, Science, and Technology Learning at NYSCI. Dr. Ba has 20 years of experience investigating children's development of digital literacy skills and the impact of innovative science, technology, engineering, and mathematics (STEM) programs in formal as well as informal educational settings. In the last seven years, his work has focused on conceptualizing, developing, and studying education programs across multiple learning environments that integrate embodied play, technology, and science learning in order to broaden STEM participation in education systems serving underrepresented students. He has served as the principal investigator on multiple National Science Foundation (NSF) and US Department of Education STEM projects. He has an interdisciplinary educational background in the social sciences and humanities with a special focus on children's exploration of physical and social environments and learning across multiple settings.

Elham Beheshti is a research associate at NYSCI where she studies how technology can enhance learning experiences that happen in museum settings. In particular, Dr. Beheshti's research focuses on creating technology-based learning tools that support deep and open-ended inquiry activities in collaborative settings. She also studies and develops learning materials in support of computing education and computational literacy. She completed her doctorate in computer science at Northwestern University.

Dorothy Bennett currently serves as director of creative pedagogy at NYSCI, responsible for developing and implementing new initiatives that reflect NYSCI's core pedagogical approach of Design Make Play. Drawing on 30 years of research and development experience in informal and formal education, Ms. Bennett helps translate this approach into practice by creating professional development experiences for NYSCI's young museum facilitators and K–12 educators, developing apps to stimulate STEM learning beyond the museum's walls, and informing the design of exhibit and program experiences to inspire a diverse audience of learners. Prior to joining NYSCI, she conducted NSF-funded foundational research in gender equity and design-based STEM education through Education Development Center, Inc's (EDC) Center for Children and Technology, Bank Street College of Education, and Sesame Workshop, collaborating with national and international partners such as IBM, the Australian Children's Television Foundation, higher education schools of engineering, and K–12 educators nationwide.

Marcia Bueno is director of NYSCI Neighbors at NYSCI. Ms. Bueno first joined NYSCI through the Science Career Ladder program over 15 years ago, while pursuing her BS in computer science at NYU Polytechnic Institute. She has previously worked in NYSCI's Alan J. Friedman Center for the Development of Young Scientists, most recently as manager of Explainers. In her roles, she strives to develop accessible and relevant opportunities for communities as an entryway to exploring STEM that acknowledges community assets and experience.

Tara Chudoba is manager of professional development at the Wildlife Conservation Society (WCS). Ms. Chudoba manages all aspects of the Advanced Inquiry Program (AIP), a ground-breaking online master's degree that combines web-based courses through Miami University with face-to-face experiential and field study at WCS. Prior to her work at WCS, she had over 12 years of programmatic, research, and development experience in museums and informal education spaces, including managing STEM education and outreach efforts at NYSCI. She strives to always work on building community and cultivating and fostering relationships in every aspect of her personal and professional life.

Katherine McMillan Culp is the chief learning officer at NYSCI where she oversees research, exhibit, and program development; educational and community outreach; and youth development programs. Dr. Culp was previously a senior research scientist at EDC and has held leadership roles on numerous research and development projects funded by NSF and the Institute of Education Sciences at the US Department of Education, including many projects exploring new applications of digital media and technologies to support middle-grade learners. Her current research projects leverage and inform NYSCI's efforts to create equitable informal STEM learning experiences that emphasize the interests and agency of NYSCI's highly diverse audiences across the lifespan. She is a Phi Beta Kappa graduate of Amherst College and holds a PhD in developmental psychology from Teachers College, Columbia University.

Xiomara Flowers is presently a bilingual specialist in the Uniondale School District of Long Island, New York. Ms. Flowers was one of the lead educators who helped co-create and facilitate the Digital Design for English Language Learners (ELLs) program sessions. She is currently part of a dynamic team of five educators at her elementary school piloting a special multi-year program called Dream Keepers and Dream Chasers for ELLs to help them develop sustained relationships with school staff and to foster continuous academic achievement throughout their elementary school experiences. She has spearheaded the bilingual education program in the Uniondale School District for the past 25 years and is an adjunct professor at Molloy College and Queens College, teaching bilingual and multilingual education courses that focus on the importance of understanding the role of culture in ELLs' instruction. Ms. Flowers was also a founding member and academic director of the New Hope Academy Charter School in Brooklyn, New York.

Andrés Henríquez is vice-president of STEM learning in communities at NYSCI, leading the NYSCI Neighbors Initiative. Mr. Henríquez brings a broad expertise to this position, having previously worked as a program officer at the National Science Foundation and the Carnegie Corporation of New York. At Carnegie he led the education division's work in establishing the field of adolescent literacy and oversaw the work in college and career-ready standards and assessments, which included the writing and adoption of the Next Generation Science Standards. Earlier in his career at the Center for Children and Technology, he was part of the community transformation in Union City, NJ, where he led a partnership between Bell Atlantic and the Union City Schools that received national recognition from President Clinton and Vice President Gore. He serves as a board member of Excelencia in Education and is a former alumni trustee of Hamilton College. Mr. Henríquez received his undergraduate degree from Hamilton College and a master's degree from Columbia University.

Margaret Honey is president and CEO of NYSCI. Dr. Honey is deeply committed to using the museum as a platform through which it can nurture a generation of creative and collaborative problem solvers in STEM fields. Under her leadership, NYSCI has developed its Design Make Play approach to learning, which supports STEM experiences that provide learners with opportunities to construct knowledge, build on prior experiences, and investigate personally and socially meaningful problems. Dr. Honey is a graduate of Hampshire College and holds a doctorate in developmental psychology from Columbia University. She has shared lessons learned before Congress, state legislatures, and federal panels, and through numerous articles, chapters, and books. She is a member of the National Academy of Education and currently serves as a member of the Division of Behavioral and Social Sciences and Education Advisory Committee for the National Academies of Sciences, Engineering, and Medicine. She also serves on the boards of Bank Street College of Education, the Scratch Foundation, and Post University.

Danny Kirk is manager of Maker Space programs at NYSCI where he supports an amazing team of facilitators as they design and implement creative maker-centered programming. Mr. Kirk is a longtime youth educator and program facilitator with a passion for making and hands-on learning. He has a BA in history from UC Berkeley with a focus in education and social movements in modern Latin America. Before joining the NYSCI team, he was the education community coordinator at Maker Ed in Oakland.

Susan M. Letourneau is a senior research associate at NYSCI. She collaborates with educators and designers to develop and study museum experiences that emphasize play, exploration, and creative expression as avenues for STEM learning for children and their families. She has worked in children's museums, science centers, and psychology laboratories for over ten years, including a collaborative research position with Brown University and Providence Children's Museum, an interdisciplinary postdoctoral fellowship in the education sciences at the Graduate Center of CUNY, and a fellowship with the Living Laboratory at the Museum of Science, Boston. She holds a PhD in cognitive neuroscience from Brandeis University.

C. James Liu is a senior research associate at NYSCI. Dr. Lui's work focuses on learning motivation in informal education. His recent projects include developing and evaluating STEM-related programs, curricula, and activities for children and their families, and conducting research on museum educators and their professional development. He holds a PhD in educational psychology from Purdue University and two master's degrees from the City College of New York (CCNY) in museum studies and psychology.

Leilah Lyons is director of digital learning research at NYSCI and an adjunct research associate professor in computer science and the learning sciences at the University of Illinois at Chicago. For the last 13 years, Dr. Lyons has been engaged in designing, building, and studying how technology can be used to support groups of learners in museums. She has tested her technological interventions in museums like the Exploratorium in San Francisco, the Ann Arbor Hands-On Museum in Michigan, the Brookfield Zoo and the Jane Addams Hull-House in Chicago, and NYSCI. She received her doctorate in computer science (with a concentration in intelligent systems) and graduate certificate in museum studies from the University of Michigan.

Aditi Mallavarapu is a computer science doctoral candidate at the University of Illinois at Chicago and a research assistant at NYSCI. Her research projects have the shared goal of collaborating with practitioners to build computational methods and analytical tools to support and improve exploration-based learning.

Kate Maschak is an experienced instructor in both formal and informal education, teaching K–6 grades, specifically in special education. Her recent works at NYSCI include professional development activities for elementary school teachers, hands-on environmental education workshops for high school students and teachers, literacy and engineering curriculum activities, digital engagement and algorithmic thinking experiences, and family workshops on data science.

Amelia Merker is an educator, artist, and designer. She started at NYSCI in 2015 as an Explainer while completing her education degree at Queens College. She transitioned into working in Design Lab, first as a resident and then as a coordinator in 2016. In her time in Design Lab, she focused on creating opportunities for visitors and Explainers to build confidence and have agency in their learning. Currently, Ms. Merker project manages the exhibits team at NYSCI, helping new exhibitions come to the museum floor. She holds a BA in fine arts with a concentration in installations and sculpture, as well as a BA in psychology and education from Queens College. In her free time, she enjoys traveling, painting, and the great outdoors.

Delia Meza is early childhood education manager at NYSCI. Working with teachers, families, and students, Ms. Meza has had a key role in developing NYSCI's early childhood initiatives, emphasizing early childhood programming best practices and sensory-sensitive learning. She holds an MA in leadership in museum education from Bank Street College of Education.

Priya Mohabir is the senior vice president of youth development and museum culture at NYSCI. As the founding director of the Alan J. Friedman Center for the Development of Young Scientists, Ms. Mohabir has enhanced the vision of our youth development programs, as she leads the development, management, and execution of the Friedman Center's youth and workforce development programs and activities. A former participant of these programs, she has been with the museum for the last 20 years and has built a program culture that recognizes and integrates Explainer expertise in the ways we shape NYSCI experiences.

Satbir Multani is exhibits manager at NYSCI. She started working at NYSCI as an Explainer in 2008 through a teacher certification partnership with CCNY. Since then, she has transitioned through the Science Career Ladder, worked on Design Lab since its inception, and is currently developing exhibitions cross departmentally and with the local community. Ms. Multani has a BA in philosophy from CCNY, with minors in biology, chemistry, and education, and an MPS from the Interactive Telecommunications Program at NYU Tisch.

Anthony Negron is the manager of digital programming at NYSCI and has been with NYSCI since 2004, when he started as an Explainer right out of high school. Mr. Negron manages several initiatives that utilize digital tools and software to educate K–12 youth on topics covering a vast array of STEM concepts and practices. He is also a key member within the Hive NYC Learning Network, which is a citywide laboratory where educators, technologists, and mentors design innovative, connected educational experiences for youth. His work with youth has been presented at conferences such as the Association of Science-Technology Centers, MozFest, Maker Faire NYC, and the International Society for Technology in Education.

Christina O'Malley is a New York State Master Teacher. Ms. O'Malley began her teaching career in 2006 at her alma mater, Akron Central, teaching high school technology and engineering. Her passion resides in coaching middle school students in building creative, critical, and self-regulated thinking skills to create solutions to tomorrow's problems; STEM is the perfect canvas for practicing this. In her Creative Design & Engineering class, students learn science, math, and technological skills to collaboratively engineer solutions to design challenges. She is an active learner and facilitator of the New York State Association for Computers and Technologies in Education and the International Society for Technology Education. She is an engaged member of the Akron Rotary Club and adviser to the high school Interact Club; both of these service organizations take action to create lasting change in her community and globally. She earned her BS in technology education from Buffalo State and an MS in differentiated instruction with a gifted education extension from Canisius College.

Dana Schloss is NYSCI's director of exhibit experiences with expertise in designing, developing, and testing experiences that help visitors take creative risks, build confidence and competence, question assumptions, and explore new ways of understanding the world. Dana has an MFA in museum exhibition design from University of the Arts ('06) and has spent time at museums all over North America. They helped develop exhibits to build a new science center at Telus Spark in Calgary, and they were a tour guide and maintainer at the Museum of the Moving Image in Queens. They ran the Making and Tinkering workshops and served as the manager of museum experience at ScienceWorks, a tiny rural museum in Oregon.

Heidi Slouffman is a fifth-grade mathematics teacher in New York City. Over the last 23 years, Ms. Slouffman has served communities as an educator in the Bronx, Brooklyn, and Queens, teaching upper elementary and middle school children to incorporate English into their everyday studies and stoking their curiosity for the world around them. With her master's degree and certification in teaching English to speakers of other languages, she loves advocating for New Yorkers, immigrant families, and all children's educational needs. She counts her experience at NYSCI as one of her favorite learning experiences to date. As an educator, her proudest moments unfold daily as she watches the love of learning grow in the creative sixth-grader and imaginative kindergartener who calls her Mom.

Daisy Tetecatl is a first-generation, Mexican American graduate of the Herbert H. Lehman College of CUNY with a bachelor's degree in biology. Ms. Tetecatl is a Corona, Queens native and has been a participant in NYSCI's signature Science Career Ladder for seven years, engaging with the museum's visitors from her own community and all over. She is currently at the Senior Explainer rank, serving as a resource and mentor for the professional development of youth in the program. Due to her passion for STEM and enjoyment of working with people of all backgrounds, she intends to pursue a career in medicine in the near future.

Wren Thompson is a behavioral researcher, previously a research assistant at NYSCI, where she studied how diverse audiences understood scientific concepts presented within museum and technology contexts. Her degree in wildlife biology and experience in science education sparked her research interests in understanding how audiences learn science in informal spaces. Ms. Thompson's formative data collection and analysis efforts informed the design of NYSCI's Beyond the Walls efforts, including educational applications like The Pack and Playground Physics.

Samantha Tumolo is the coordinator of weekday maker programs at NYSCI. Ms. Tumolo has a strong interest in the intersection of art and science and explores ways to design programs that balance the two. She also works directly with NYSCI's community engagement team to deliver inclusive educational experiences to NYSCI's local underserved community. It is her goal to design experiences that welcome all types of visitors and immerse them in a flow state, leaving them excited and driven to make more. She received her bachelor's degree in environmental studies from Temple University and also participated in the TUteach program focused on STEM education.

Laycca Umer is manager of research, exhibits, and programs at NYSCI. Her work over the past ten years has focused on developing inclusive and equitable learning opportunities, designing innovative digital learning tools using research-based approaches, and refining organizational systems. Ms. Umer strives to bridge the gap between the playful museum experience and traditional classroom learning and support the growth of positive workplace culture. She holds a BSEd in childhood education and psychology and an MS in TESOL from CUNY City College.

Stephen Uzzo is chief scientist for NYSCI. Dr. Uzzo develops and leads research initiatives to integrate cutting-edge science into teaching and learning. Through his work, he cultivates communities of practice in complexity, data-driven science and engineering, and STEM literacy of the public. His background includes over 30 years of experience in the research of connected systems and teaching and learning in STEM, and prior to that, ten years in video and computer graphics systems engineering and distance learning. He holds a terminal degree in network theory and environmental studies and serves on a number of institutional and advisory boards related to his interests.

about the contributors

Betty Wallingford served as a research assistant at NYSCI, where she assisted on multiple grant-funded research projects in the learning sciences and exhibit design. Her work focuses on how open-ended and hands-on learning experiences can contribute to children's confidence as learners and makers. She is now working towards an MFA in child culture design at the University of Gothenburg in Sweden.

David Wells is director of maker programming at NYSCI. Mr. Wells oversees the design and implementation of maker-related programs and works with formal and informal educators to create and implement maker activities and maker spaces in their setting. He has worked on a wide variety of projects, including designing exhibits for the early childhood space, developing emergent curriculum for educational programming, and creating educational videos for teachers and students. A self-proclaimed "maker of things," Mr. Wells also designs site-specific interactive art installations using discarded technology, audio experimentation, and digital media to create an interesting yet whimsical experience for his viewers. He received his BFA from the Fashion Institute of Technology and a master's in museum education from Bank Street.

Acknowledgments

We would like to begin by acknowledging that this book was born in the midst of the global COVID-19 pandemic and the equally disturbing pandemic of American racism and the brutal deaths of far too many African Americans. At the New York Hall of Science (NYSCI), we are deeply committed to an approach to science, technology, engineering, and mathematics (STEM) learning – Design Make Play – that has equity and inclusion baked into its very core. This is not a remedy for our country's long history of systemic racism, but a strategy we can enact to ensure that our work on STEM learning creates a broad and inclusive umbrella of engagement. For those of us involved in this project, COVID-19 has, ironically, provided what we have come to call "the gift of time." We were able to turn our attention to writing, and to create a product that we hope will be of value to those of us who work in broadening STEM learning. This book, and the trajectory of work it presents, would not have become a reality without this gift of time and without the contributions of NYSCI's many supporters.

First and foremost, we thank the entire NYSCI team – all of our colleagues who contribute to realizing NYSCI's mission and work tirelessly to engage all of our audiences. We also thank the chapter authors for their dedication, collaboration, patience, and prompt responses to our endless queries: Yessenia Argudo, Harouna Ba, Elham Beheshti, Dorothy Bennett, Marcia Bueno, Tara Chudoba, Xiomara Flowers, Andrés Henríquez, Danny Kirk, Susan M. Letourneau, C. James Liu, Leilah Lyons, Aditi Mallavarapu, Kate Maschak, Amelia Merker, Delia Meza, Priya Mohabir, Satbir Multani, Anthony Negron, Christina O'Malley, Dana Schloss, Heidi Slouffman, Daisy Tetecatl, Wren Thompson, Samantha Tumolo, Laycca Umer, Stephen Uzzo, Betty Wallingford, and David Wells.

We also wish to recognize NYSCI's Board of Trustees, whose incredible generosity and passion for this institution enables us to realize our impact well beyond the walls of the museum. To all of our supporters – NYSCI's President's Council members, foundations and corporations, and the government funding agencies that support much of our research work – we extend our deepest thanks. We want to especially acknowledge the support of the Gordon and Betty Moore Foundation, who got behind our first Design Make Play publication in 2013 and supported us in bringing this most recent book to light.

We also owe a debt of gratitude to our many partners – educators and school administrators from New York City and across New York State, families and their children from Corona, Queens, and our loyal visitors from the greater New York City metropolitan area. We have the utmost respect for all of you and extend our thanks for your input, guidance, and willingness to experiment and innovate in the service of creating powerful STEM experiences for all learners. It is you who make it possible for us to combine research and practice and address issues of agency, inclusion, and equity in creative STEM learning experiences.

We are also honored to be part of the broader science center and informal learning community. We are by no means alone in the work we are doing, and we value the generosity and guidance that comes through our collaborations with colleagues who work in these dedicated and creative organizations.

We extend our deepest gratitude to Nancy Kober, a talented writer and editor, for her valuable contributions to the writing of this book. She worked tirelessly and flawlessly to provide insightful guidance to all of the authors. Thanks also to our NYSCI colleague Danny Loi for designing the cover page.

Finally, we appreciate the dedication of our editors at Routledge, Simon Jacobs and AnnaMary Goodall, who embraced this project from its inception.

Introduction
Design Make Play for Equity, Inclusion, and Agency

Harouna Ba, Katherine McMillan Culp, and Margaret Honey

Origins of This Book

In 2013 the New York Hall of Science (NYSCI) published our first effort to define and illustrate our vision for science, technology, engineering, and mathematics (STEM) learning – *Design, Make, Play: Growing the Next Generation of STEM Innovators* (Honey & Kanter, 2013). That book addressed the urgency of encouraging the development of STEM innovators of the future and asked the field to pay special attention to a broad approach to STEM learning across both formal and informal environments. Most of the contributions in the first Design Make Play book were framed by and heavily influenced by the emerging maker culture and do-it-yourself ethos (Dougherty, 2013; Swan, 2014; Martinez & Stager, 2013; Wilkinson & Petrich, 2014), the federal government's consideration of innovative STEM education as a top contributor to the country's future economic well-being (Kalil, 2013), and the release of the *Framework for K-12 Science Education* (National Research Council [NRC], 2012). While we think the work showcased in that book did make real contributions to the field, in hindsight it was a relatively simple task to present a broad aspirational definition of Design Make Play and describe its first maker and design instantiations in predominantly informal learning environments.

The question we have wrestled with more recently is how Design Make Play can be advanced well and deeply within the culture and the practical realities of a functioning, public-facing cultural institution like NYSCI. Figuring out what Design Make Play means across our work at NYSCI, and how to bring that meaning to life in a busy institution serving many audiences with many different needs, has been a process of experimentation and iteration. Program developers, facilitators, educators, exhibit designers, researchers, and external partners have all contributed to the process of interpreting and reinterpreting how to realize this aspirational approach to engagement and learning through our collective work.

With extensive funding from the National Science Foundation, US Department of Education Institute of Education Sciences, Institute of Museum and Library Sciences, National Institutes of Health, and corporate and private philanthropies, our understanding of Design Make Play has been clarified and made more precise, through practice and discussion but also through the empirical rigor of innovative research and development projects carried out in recent years. We now understand that Design Make Play encompasses the following empirically tested ideas about STEM teaching and learning:

- Sophisticated scientific reasoning can emerge even at very young ages when it is elicited and supported.
- Environments rich in materials and language invite exploration, observation, question-asking, and discussion, all of which are critical contributors to future learning.
- Understanding difficult concepts requires multiple and varied opportunities to ask questions, try out solutions, and reflect on emergent knowledge.
- Interaction with skilled facilitators who ask good questions and provide feedback and guidance enriches learning.

At its core, Design Make Play is a constructivist pedagogy that invites broad STEM participation by making engagement and learning irresistible (NRC, 2009; NRC, 2000). Design Make Play experiences invite learners to play with and explore a variety of materials and physical experiences and to ask questions, take risks, recognize and pursue their curiosity, create and share compelling narratives that spur new questions, and experience success. The concept of *design* emphasizes intentionality, purpose, and agency in problem-solving for generating divergent solutions. Learners define problems, test them, and iterate on them in the process of arriving at potential solutions or questions (Li et al., 2019). *Make* refers to both embodied and digital experiences and interactions with materials, tools, and processes and nurtures the building of skills and creation of physical objects to represent ideas and the world around us. As learners gain a sense of control and power over their environment, they develop a sense of agency and efficacy (Bevan et al., 2015; Vossoughi & Bevan, 2014). *Play* reminds us of the importance of structuring learning in ways that can be continually remade and modified by learners, and of linking learning to physical action and creative thought experiments exploring complex scientific ideas (Bergen, 2009). Playful activities are designed to support intrinsic motivation, deep engagement, and deep learning (Fine, 2014; Mehta & Fine, 2019) and are effective mechanisms for

harouna ba et al.

encouraging creativity and facilitating innovation (Bateson & Martin, 2013). We seek to foster the excitement of self-directed exploration and tap into the joy of learning that is intrinsic to young people's play.

Design Make Play encapsulates the strategies we use to help learners experience their agency and creativity as learners and makers, master complex concepts and phenomena, and discover how they can put the tools and perspectives of the STEM disciplines to work to address their own concerns and questions. When we design experiences that reflect the aspirations of the Design Make Play approach, we utilize five core principles that support equity, inclusion, and learners' agency:

1. *Putting learners and play at the center* leverages learners' natural instincts to engage playfully with compelling ideas and materials.
2. *Positioning learners as creators* engages learners as producers and makers, not just consumers, of content, materials, and material objects. It can propel students to build new knowledge through the creative application of their skills and ingenuity.
3. *Tackling problems that one thinks are worth solving* supports interest-driven problem-solving in which learners actively define and shape the tasks they are working on.
4. *Cultivating divergent solutions* presents problems that that can be solved in multiple ways, inviting learners to apply a wide variety of skills and knowledge in finding a solution.
5. *Providing an accessible invitation* offers activities with a low barrier to entry, a high ceiling of potential complexity, and a wide berth for the creative expression of ideas.

Themes of This Book

In this new book, *Design Make Play for Equity, Inclusion, and Agency*, we capture the ways in which current work at NYSCI is evolving to best serve our audiences and extend to formal education and community settings. Our work is situated in a context informed by three important developments that are themselves dynamic and evolving: (a) our rapidly growing understanding of how people learn in both informal and formal learning environments (National Academies of Sciences, Engineering, and Medicine [NASEM], 2018; NRC, 2009; NRC, 2000); (b) the urgent need for richer, more ambitious STEM learning opportunities for non-dominant learners (African Americans, Latinx, women, immigrant communities, and people with disabilities) who have been underserved by the K–12 education system (NASEM, 2018) and who have often been left out of shaping the cultures and practices of the STEM disciplines themselves; and (c) the increasing complexity of current STEM research and innovation, which is swiftly expanding and dramatically affecting the daily lives of the public (American Institutes for Research, 2016). Together, these developments require science centers and museums to reconsider our approaches to STEM education and to seek out more ambitious and more equitable forms of engagement that are also appealing and accessible to a diverse range of audiences (Dawson, 2019; Feinstein, 2017).

Three broad themes emerge across these chapters: designing to promote visitors' agency in informal settings, reorganizing museums for visitors' full participation and contribution, and creating experiences grounded in embodied and playful activities for productive engagement and deeper learning in formal settings. Importantly, each chapter reflects the commitment of our colleagues to share the lessons they are learning as we work to design STEM learning experiences that are truly equitable and inclusive and that foster learners' agency. The book is divided into three parts that elaborate on these themes, as explained below.

Part I – Designing for Visitors' Agency

Research has already made the theoretical and empirical cases that trusting visitors to pursue their own pathways through museums and make discoveries that are meaningful to them creates powerful motivation for their future learning. One strand of this work has focused on how to design innovative ways to support visitors' discovery of often complex but foundational scientific concepts (Falk, 2016; Falk & Dierking, 2010, 2016; Gutwill, 2008; Humphrey et al., 2005; Szechter & Carey, 2009). This work has helped to move science centers and museums away from covering content and toward encouraging visitor-driven and open-ended exploration. Audience research has also made the case that visitors to science museums are most motivated to learn when they can engage with new ideas in the context of their own cultural identities, experiences, and perspectives (Golding & Modest, 2013). In addition, work on public engagement with science has argued for the value of two-way engagement and discussion between experts and publics, de-emphasizing the delivery of authoritative scientific knowledge in favor of "opportunities for an exchange of knowledge, ideas, and perspectives that involves the participation of all aspects of society – publics, scientists, and decision makers" (McCallie et al., 2009, p. 11). All of this work emphasizes the importance of mutual learning, of exposing and cultivating our recognition of the STEM disciplines as cultural practices and the integration of STEM-based inquiry and cultural values into decision making and problem-solving. It also depends upon learners' readiness to take up opportunities to speak up, pose questions, and make active contributions to their own learning and to collective learning experiences. In turn, we have recognized that not all visitors are equally comfortable with this proposition, and many may need both other pathways into these experiences and low-risk opportunities to practice drawing on their own agency as learners to contribute to shared informal learning experiences.

Four chapters of the book build on this body of work to further explore the ways in which museums can design programs and exhibits that invite, cultivate, and support visitor agency. Chapter 1 presents an overview of existing research exploring the construct of agency in informal learning environments. Chapter 2 discusses the role that floor facilitators, whom we call Explainers, can play in cultivating visitors' agency and highlights the strategies that NYSCI has used to redesign its Explainer training program. Chapter 3 examines the role that narrative and empathy can play in engineering design problems to support more engagement and persistence among girls. Chapter 4 looks at the role that participatory design can play in building relational agency for floor facilitation staff. As a whole, these chapters describe how NYSCI is shifting our efforts away from acting as content experts and toward facilitators of visitors' own pathways as learners, responding and contributing to exhibits and to one another.

Part II – Relinquishing Power and Authority in Informal Settings

Just as we seek to invite and support visitors' agency as learners, we also have to learn how to reframe the science museum itself. We have been exploring how to reimagine NYSCI not as a repository of elite knowledge but as a place where visitors can draw on their own personal experience and expertise to engage with, reflect on, and contribute to the everyday work of scientific discovery, the engineering of our everyday environment, and the evolution and critical examination of technological innovations. Models for this kind of reframing come from the work of museum leaders and exhibit designers including Simon (2010, 2016) and Mortati (2014), among others (Rovatti-Leonard, 2014; Villeneuve & Love, 2017; Murawski, 2018). These authors often describe their goals in terms of creating participatory museums (Simon, 2010). The intersection of genuine community participation with the complexities of the STEM disciplines is territory we and others are just beginning to explore, and this raises different challenges from those in history and cultural museums where staff are charged with working more intensively and personally with the community they serve and where community wisdom dovetails powerfully with the historical narratives the museum may house (see, for example, Iervolino, 2013).

Several chapters of this book examine how NYSCI is designing a range of exhibits and programmatic experiences to more deeply engage the diverse audiences we serve. Chapter 5 describes NYSCI's work to support more family STEM learning among residents of our neighboring first-generation community of Corona, Queens. Chapter 6 presents NYSCI's Museum Makers program, a workshop series that involved families in data modeling processes through exhibition design. Chapter 7 explores strategies we have used to open up the museum's maker programs to families who typically have not been invited into maker cultures. Chapter 8 discusses NYSCI's Innovation Institute, a program designed to get teenagers solving problems in their local communities through design and making activities at the museum. Each of these programs has explored strategies for putting ambitious tools and practices of STEM in the service of the questions and interests of learners who might not otherwise have access to these ways of framing questions, designing solutions, and creating expressions of their ideas and problem solutions. As each chapter also suggests, we are invested in pushing this paradigm further and exploring how to broaden our definition of "participatory" design and programming while also remaining a source of reliable information and inspiring ideas for the communities we serve.

Part III – Playing and Learning Across Settings

Programs created with Design Make Play principles at the core are emerging as potent methods to involve a broad constituency in STEM engagement and learning in formal settings. They offer a low barrier to entry for reluctant STEM learners and a high ceiling for achievement for active STEM learners. The activities of informal learning environments can be leveraged to inform the development of innovative science curricula, professional development, and digital tools for formal STEM learning environments. Science museums more broadly have demonstrated that there are multiple models for successfully reaching beyond our walls to engage educators who work in schools and other community settings. The remaining chapters of the book explore how NYSCI's initiatives are drawing on strategies that have been incubated in an informal context to extend our reach into formal education settings. Chapter 9 focuses on the use of multimodal learning strategies that support the active involvement of English language learners in solving mathematical problems. Chapter 10 presents research on the power of play and embodied strategies to support productive engagement for physics learning in school settings. Chapter 11 discusses work that involves integrating computational thinking strategies into an elementary school's curriculum. Finally, Chapter 12 explores how to meaningfully integrate computational thinking in a STEM learning context to solve environmental challenges and discusses how to place more continuous agency in the hands of learners.

Where We've Been and How We Hope This Book Will Contribute to the Field of STEM Learning

Many of us who study teaching and learning make the mistake of thinking ahistorically. We suffer from the conceit that our insights are brand new inventions that will surely lead to transformative revolutions in practice and policy. Too often, the incentives that drive our work encourage this thinking. However, bringing together the chapters in this book and revisiting all the intersections and histories behind these projects reminded us, as the editors, that all of our work originates within a complex web of dialogues, relationships and institutions that have shaped the richer story of our motivations. This web has also shaped our contributions to the shared, ongoing effort to ensure that everyone can engage in meaningful, powerful STEM learning experiences and to broaden ideas about whom the STEM disciplines are for and how STEM resources are deployed. Recognizing those through-lines and intersections that have evolved over time and influenced our perspectives and our thinking is essential to understanding why all of us do what we do.

As editors of this volume, the three of us have been colleagues for more than 30 years. We had the good fortune of coming together in the late 1980s at the Center for Children and Technology (CCT), an innovative research and development group founded in 1980 at Bank Street College of Education in New York City (Pea, 2016). CCT and Bank Street were significant sources of inspiration for each of us, molding a set of core learning values that have remained consistent throughout our years together. Anchored in a progressive pedagogy defined by Lucy Sprague Mitchell, Bank Street's founder, and John Dewey, the 20th-century education philosopher, the institution propagated an overarching approach to learning that guides us to this day: a belief that young people learn through doing and through engaging in tasks that have personal relevance to them and build on their prior understandings (Antler, 1987; Dewey, 1938/1997, 1902/1956). These beliefs have since been backed up by many decades of research in the learning sciences (NASEM, 2018; NRC, 2000, 2009). We continue to benefit from the maturation of the learning sciences as a discipline and from the progress of our colleagues in that field, especially from their ongoing efforts to reimagine science learning in more culturally responsive ways.

Each of us came to NYSCI at a significant moment in the museum's evolution. Margaret Honey joined as president and CEO in November 2008. She followed the 22-year tenure of Dr. Alan Friedman, a physicist turned science educator who devoted his considerable expertise to exploring how informal learning environments could be used to engage the broader public in science education. As a developmental psychologist and learning scientist, she cares deeply about ensuring that all learners have experiences that make them feel empowered – smart, competent, and successful. Her vision has been to leverage the museum as a learning lab that can create new products and programs and carry out research that will help to shape and inform the future of STEM engagement and learning.

Harouna Ba joined NYSCI five years later, in November 2013, to direct the Sara Lee Schupf Family Center for Play, Science, and Technology Learning and develop a series of programs to support science engagement and learning through play and embodied learning across learning environments. An environmental psychologist who had spent the early part of his career studying how urban environments shape children's growth and development, he has been committed to understanding how children's geographic explorations aid in the development of problem-solving, collaboration, and inquiry skills. Not coincidentally, it was Lucy Sprague Mitchell who concluded that children's everyday geographic explorations are key tools for building bridges between informal pedagogy – what she called the "curriculum of experience" – and the often more remote academic curriculum (Mitchell, 1934/1991).

Katherine McMillan Culp joined NYSCI in August 2015 as the organization's first-ever chief learning officer, a position that would signal NYSCI's commitment to integrating a cohesive approach to supporting informal STEM learning into every area of our work. A developmental psychologist by training, she brought long experience in applied research and development to NYSCI. Her experience developing research that responds to problems of policy and practice, and building collaborative relationships among researchers and practitioners, has allowed us to begin integrating and aligning our work across departments in new and productive ways. Over the past five years she has sought to surface the diverse, inspiring expertise of NYSCI's staff and work with them to imagine and create ambitious, inclusive, and scalable informal STEM learning experiences.

The three of us, and the work that our NYSCI colleagues and external partners make possible in the museum, in schools locally and nationwide, and in our local community, are deeply motivated by a desire to help all learners acquire the critical habits of mind, capacity for collaboration, and opportunities to act on the world around them that spark all forms of creativity and discovery. This vision drives all of our work at

NYSCI, defining the aspirational Design Make Play approach to STEM engagement and learning that we have tried to unpack and articulate here.

We hope that this volume will inspire our readers, just as the work of our colleagues has inspired us, especially at the end of a year that has been challenging in both unimaginable and all too familiar ways. More than ever, we recognize that the phrase Design Make Play references an ideal, a vision of inclusion, community, and self-determination for all STEM learners that motivates us but will always be a goal we are working toward. We hope that you, like us, will draw on the insights in these chapters to help you clarify your vision and your goals for supporting learners. We hope they will help you to recognize and foster those moments of truly powerful, liberating STEM learning and teaching when you are lucky enough to find them, and to push our collective efforts forward toward a more democratic vision of the STEM disciplines as resources that are at the service of all learners, doers, and visionaries.

References

American Institutes for Research. (2016). *STEM 2026: A vision for innovation in STEM education*. US Department of Education.

Antler, J. (1987). *Lucy Sprague Mitchell: The making of a modern woman*. Yale University Press.

Bateson, P., & Martin, P. (2013). *Play, playfulness, creativity and innovation*. Cambridge University Press.

Bergen, D. (2009). Play as the learning medium for future scientists, mathematicians, and engineers. *American Journal of Play, 1*(4), 413–428.

Bevan, B., Gutwill, J. P., Petrich, M., & Wilkinson, K. (2015). Learning through STEM-rich tinkering: Findings from a jointly negotiated research project taken up in practice. *Science Education, 99*(1), 98–120.

Dawson, E. (2019). *Equity, exclusion, and everyday science learning: The experiences of minoritized groups*. Routledge.

Dewey, J. (1956). *The child and the curriculum and the school and society*. University of Chicago Press. (Original work published 1902)

Dewey, J. (1997). *Experience and education*. Kappa Delta Pi. (Original work published 1938)

Dougherty, D. (2013). The maker mindset. In M. Honey & D. Kanter (Eds.). *Design, make, play: Growing the next generation of STEM innovators* (pp. 7–11). Routledge.

Falk, J. H. (2016). *Identity and the museum visitor experience*. Routledge.

Falk, J. H., & Dierking, L. D. (2010). The 95 percent solution: Schools is not where most Americans learn most of their science. *American Scientist, 98*(6), 486–493.

Falk, J. H., & Dierking, L. D. (2016). *The museum experience revisited*. Routledge.

Feinstein, N. W. (2017). Equity and the meaning of science learning: A defining challenge for science museums. Science Education, *101*(4), 533–538.

Fine, S. M. (2014). "A slow revolution": Toward a theory of intellectual playfulness in high school classrooms. *Harvard Educational Review, 84*(1), 1–23.

Golding, V., & Modest, W. (Eds.). (2013). *Museums and communities: Curators, collections and collaboration*. Bloomsbury.

Gutwill, J. P. (2008). Challenging a common assumption of hands-on exhibits: How counterintuitive phenomena can undermine inquiry. *Journal of Museum Education, 33*(2), 187–198.

Honey, M., & Kanter, D. (2013). *Design, make, play: Growing the next generation of STEM innovators.* Routledge.

Humphrey, Gutwill, & Exploratorium APE Team. (2005). *Fostering active prolonged engagement: The art of creating APE exhibits.* Left Coast Press.

Iervolino, S. (2013). *Ethnographic museums in mutation experiments with exhibitionary practices in post/colonial Europe.* [Doctoral dissertation, University of Leicester]. Semantic Scholar database. www.semanticscholar.org/paper/Ethnographic-museums-in-mutation-experiments-with-Iervolino/5e98e889c9edb1f44b8531cdab e91d3f5ab4bc11

Kalil, T. (2013). Have fun – learn something, do something, make something. In M. Honey & D. Kanter (Eds.). *Design, make, play: Growing the next generation of STEM innovators,* (pp. 12–16). Routledge.

Li, Y., Schoenfeld, A. H., diSessa, A. A., Grasser, A. C., Benson, L. C., English, L. D., & Duschl, R. A. (2019). Design and design thinking in STEM education. *Journal for STEM Education Research, 2*(2), 93–104. https://doi.org/10.1007/s41979-019-00020-z

Martinez, S. L., & Stager, G. (2013). *Invent to learn: Making, tinkering, and engineering in the classroom.* Constructing Modern Knowledge Press.

McCallie, E., Bell, L., Lohwater, T., Falk, J. H., Lehr, J. L., Lewenstein, B. V., Needham, C., & Wiehe, B. (2009). *Many experts, many audiences: Public engagement with science and informal science education.* A CAISE Inquiry Group Report. Center for Advancement of Informal Science Education (CAISE). www.informalscience.org/sites/default/files/PublicEngagementwithScience.pdf

Mehta, J., & Fine, M. (2019). *In search of deeper learning: The quest to remake the American high school.* Harvard University Press.

Mitchell, L. S. (1991). *Young geographers: How they explore the world and how they map the world* (4th ed.). Bank Street College of Education. (Original work published 1934)

Mortati, M. (2014, Spring,). Design intentionality and the art museum. *Exhibitionist,* 34–39.

Murawski, M. (2018). *Towards a more human-centered museum: Part 2. Building a culture of empathy.* https://artmuseumteaching.com/2018/02/06/towards-a-more-human-centered-museum-part-2-building-a-culture-of-empathy/

National Academies of Sciences, Engineering, and Medicine. (2018). *How people learn II: Learners, contexts, and cultures.* The National Academies Press. https://doi.org/10.17226/24783

National Research Council. (2000). *How people learn: Brain, mind, experience, and school.* The National Academies Press.

National Research Council. (2009). *Learning science in informal learning environments: People, places, and pursuits.* The National Academies Press.

National Research Council. (2012). *Education for life and work: Developing transferable knowledge and skills in the 21st century.* The National Academies Press.

Pea, R. (2016). The prehistory of the learning sciences. In M. A. Evans, M. J. Packer, & R. K. Sawyer (Eds.), *Reflections on the learning sciences*. Cambridge University Press.

Rovatti-Leonard, A. (2014). The mobile LAM (Library, Archive & Museum): New space for engagement. *Young Adult Library Services, 12*(2), 16–21.

Simon, N. (2010). *The participatory museum*. Museum 2.0.

Simon, N. (2016). *The art of relevance*. Museum 2.0.

Swan, N. (2014, July 6). The "maker movement" creates D.I.Y. revolution. *The Christian Science Monitor*. www.csmonitor.com/Technology/2014/0706/The-maker-movement-creates-D.I.Y.-revolution

Szechter, L. E., & Carey, E. J. (2009). Gravitating toward science: Parent-child interactions at a gravitational-wave observatory. *Science Education, 93*(5), 846–858.

Villeneuve, P., & Love, A. R. (Eds.). (2017). *Visitor-centered exhibitions and edu-curation in art museums*. Rowman & Littlefield.

Vossoughi, S., & Bevan, B. (2014). *Making and tinkering: A review of the literature*. Commissioned paper for the Committee on Successful Out-of-School STEM Learning: A consensus study. National Research Council. https://sites.nationalacademies.org/cs/groups/dbassesite/documents/webpage/dbasse_089888.pdf

Wilkinson, K., & Petrich, M. (2014). *The art of tinkering*. Exploratorium.

Part I
Designing for Visitors' Agency

Designing for Agency in Informal STEM Learning Environments

Susan M. Letourneau

> Motivation to learn is fostered for learners of all ages when they perceive the school or learning environment is a place where they "belong" and when the environment promotes their sense of agency and purpose.
>
> National Academies of Sciences, Engineering, and Medicine, 2018, p. 133

Decades of research in the learning sciences have demonstrated that the most powerful learning experiences happen when individuals are able to actively construct knowledge, build on their everyday experiences, and investigate personally meaningful problems (The National Academies, 2018). In informal learning environments, educational practices are based on the understanding that learning is embedded in personal, social, and cultural experiences that accumulate and evolve over time (Falk & Dierking, 2016; Rogoff et al., 2016). Designers and educators in these settings pride themselves on offering active and engaging learning experiences that are relevant to people's lives. But do these experiences specifically support visitors' sense of agency? And how might our practices shift if we focus our attention on agency as a distinct and important construct?

On a typical day in a science center, one might observe any or all of the following taking place:

- In an interactive exhibit, a visitor pushes a button to see a video or lifts flaps to read labels.
- A facilitator starts a conversation with a group of visitors by asking what they notice about a display.
- Caregivers and children build something in an engineering design space, discussing and elaborating on their plans over a sustained period of time.
- A student looks at specimens through a microscope in a biology exhibit.
- Children play a digital game together while their caregivers sit on a bench nearby to recharge.
- Visitors quietly listen to a science film in a theater.
- Teachers and chaperones guide groups of students through an exhibit.
- A student carefully observes a demonstration and raises her hand to ask a question.

Although most of these activities involve visitor participation, they may not necessarily promote visitors' agency in the learning process – if, for example, they constrain how visitors can make discoveries, limit the questions they can ask, or fail to tap visitors' own knowledge and perspectives. Educators in informal learning environments might ask themselves questions like these: Which aspects of the experiences they offer allow learners to exercise their agency? What does "agency" mean and look like in informal learning environments, and why is it uniquely important for the kinds of learning that science centers hope to cultivate?

Researchers and educators have explored the role of agency in science, technology, engineering, and mathematics (STEM) learning from many angles and at various levels, from the individual to the systemic. This work leads us to recognize the complexity of designing learning environments that center on learners' ideas while also guiding them toward specific learning goals. This chapter briefly introduces multiple lines of research that relate to agency in informal STEM learning and link agency with the creation of equitable and inclusive learning experiences. These studies provide a helpful lens for critically examining and questioning practices used in informal learning environments. As explained in this chapter, examining our practices through the lens of agency can change how we conceptualize the relationships between scientific knowledge, designed learning environments, and the learners themselves. Each line of work raises questions for practice that can inform the design of informal experiences and environments.

Who Decides? Interactivity, Agency, and Authority

Research in science education has examined agency in the context of the negotiations that take place as educators and learners define what questions to ask, how to

susan m. letourneau

Figure 1.1 Children collectively write down their individual measurements on a shared poster-size image of the exhibit Celestial Mechanics.

find answers, and how knowledge is constructed and applied (Arnold & Clarke, 2014; Damsa et al., 2010; Miller et al., 2018). These studies demonstrate that providing opportunities for learners to shape the direction and outcome of their scientific investigations allows them to develop STEM practices in the context of their prior knowledge, perspectives, and skills (Barton & Tan, 2010; Stroupe, 2014). However, supporting agency in this way requires transforming pedagogical practices to give learners greater authority to decide what is worth knowing and how knowledge is produced (Stroupe et al., 2018). Learners' own knowledge and perspectives are seen as relevant and necessary to the learning process (see Figure 1.1). This view aligns with research conducted in informal learning environments that describes learning as an active, personal, and experiential process, mediated by people's social interactions and cultural contexts (Hein, 1998, 2006; Falk & Dierking, 2016; Rogoff et al., 2016).

Limitations on Visitors' Agency

On the surface, many informal learning experiences provide opportunities for learners to be active participants. Science centers are designed to invite free choice exploration. Visitors navigate museums based on their existing interests and agendas, making sense of scientific content presented by museum exhibits and actively seeking out connections to their lives (Ash, 2004; Ellenbogen et al., 2004; Falk & Dierking, 2016; Zimmerman et al., 2010). As illustrated in the examples at the start of this chapter, experiences in museums and science centers invite and encourage participation in many learning experiences. However, the desire to teach science in an active and individualized way is sometimes at odds with the desire to communicate established scientific knowledge. This latter approach of conveying knowledge is often what visitors have come to expect from science and what educators and designers are used to providing. For example, exhibits often include physical and multimedia interactives (buttons to press, flaps to lift, levers to pull) that prompt visitors to seek out information, but these elements may be used to communicate "correct" answers in one direction, from "experts" to learners (Whitcomb, 2006). Although the implicit messages conveyed by these modes of interaction conflict with what research has shown about the kinds of contextualized and personal experiences that support deeper learning, they may align with visitors' (and perhaps educators') expectations that science can offer the certainty of undisputed factual knowledge.

Other exhibits in informal settings focus on discovery. They provide opportunities for visitors to ask questions and build on their own experiences while exploring a scientific concept (Hein, 2006). Pioneering work at the Exploratorium, for instance, showed that designing exhibits to provide accessible entry points and multiple pathways through an experience can support active, prolonged engagement and extend visitors' exploration of scientific phenomena (Allen, 2004; Humphrey & Gutwill, 2005). With these exhibits, the goal is for visitors to explore in ways that support their understanding of accepted scientific knowledge, while giving them some control over the path they take to get there. Although these strategies are effective in deepening exploration, scientific knowledge is still framed as something external that the visitor can acquire or discover. These exhibits are often designed to ensure that visitors make discoveries in predictable ways that reinforce established knowledge (Whitcomb, 2006).

By limiting the ways in which visitors can make discoveries and by offering predetermined, authoritative explanations as the end point of the experience, museums often retain control over the questions being asked and the pathways that visitors can take to answer them. These practices can put visitors in the position of being passive receivers of information. And while this is a risk for all visitors, there is evidence that these practices disproportionately affect learners from non-dominant communities by centering on white, masculine, middle-class views of STEM and excluding the perspectives and knowledge that visitors bring (Buechley, 2013; Dawson, 2017, 2019; Gaskins, 2008). As a result, learners with many other backgrounds have to do extra work to figure out how to use exhibits and relate the information presented to their own experiences and prior knowledge (Ash & Lombana, 2011; Aikenhead, 2006; Dawson, 2014; Medin & Bang, 2014).

More Equitable Approaches

Research suggests that more equitable approaches do not limit learners' involvement to constructing previously accepted knowledge. Instead, these approaches involve learners in deciding what to learn and how the learning process should unfold (Russ & Berland,

susan m. letourneau

2019; Stroupe, 2014). This point of view aligns with the constructivist and participatory educational approaches used in many museums, which remove the authoritative voice entirely in order to allow for a dialogue in which visitors see their own experiences and ideas represented (McLean & Pollock, 2007; Whitcomb, 2006). For example, maker spaces place the artifacts that visitors create at the center of the experience (Sheridan et al., 2014), and participatory practices used in a variety of cultural institutions invite visitors to interact with one another and contribute to exhibit installations that evolve over time (Simon, 2010). In these environments, learning is viewed as a process of making meaning that cannot exist separately from the larger context of learners' everyday lives, cultures, and prior experiences (Hein, 2006).

For example, in the programs described in Chapter 7 of this volume, staff in the New York Hall of Science's (NYSCI) Maker Space created opportunities for families who visited NYSCI regularly to learn how to use a variety of new tools. Facilitators showed children how to use the tools but left the substance of their design projects entirely up to them. As families returned over time, children were able to build skills and pursue projects that interested them. Rather than simply receiving information from others or following instructions, children went on to share what they had learned with other children and their caregivers and even suggested new project ideas to museum staff. These kinds of experiences with making and design can allow individuals to shift fluidly from novice to expert as they become part of a community of learners (Gutierrez et al., 2014; Sheridan et al., 2014).

Of course, educators and designers may have concerns about relinquishing control over what and how people learn. There are inherent tensions between welcoming learners' ideas and teaching them the established knowledge and practices within scientific disciplines (Russ & Berland, 2019; Penuel & Furtak, 2019). As trusted sources of information, science centers have a responsibility to raise public awareness about scientific topics and communicate accurate information (Dilenschneider, 2017; McCallie et al., 2009). Individual exhibitions and programs aim to achieve specific learning goals that are grounded in existing scientific understanding and educational standards. Individualized, visitor-centered experiences, although engaging, can sometimes obscure these larger messages (Whitcomb, 2006).

Some guidance and constraints are also necessary to create effective visitor experiences. Completely open-ended experiences risk welcoming only those who already feel comfortable directing their own learning experiences, which could alienate newcomers and those who are less familiar with informal learning spaces or the topics being presented (Dawson, 2014; Rogoff et al., 2016). In many cases, making STEM concepts explicit and calling attention to the connections between learners' ideas and accepted scientific practices can be a sign of respect and inclusion that expands learners' perceptions of STEM and their place in it (Nasir et al., 2006; Vossoughi et al., 2013).

Emily Dawson (2014) argues that science centers can navigate these apparently competing demands. They can become more inclusive by using strategies that tap into the knowledge and experiences of a broader audience of learners in order to go beyond historically dominant views of STEM, while providing enough information for learners to engage with STEM topics in ways that are meaningful to them.

One example of this approach is a recent NYSCI exhibit about microbes, called Small Discoveries. This exhibit included the familiar reference of an ant to illustrate the scale of common microbes. Visitors were provided with hand lenses, video microscopes, and other tools to encourage them to explore samples from everyday surfaces like doorknobs, kitchen sponges, and the bottom of a shoe. This reimagined experience prompted richer conversations than the previous version of this exhibit, which communicated scientific information through explanatory text panels and other traditional modes. In the revised version, visitors of all ages asked questions, made observations, drew conclusions, and shared discoveries with each other. The scientific content served as a starting point for open-ended exploration and inquiry. The scientific accuracy of visitors' observations was secondary to the curiosity that was evoked by revealing a previously invisible microscopic world and giving visitors tools to investigate it.

Questions for Practice

In order to move toward more inclusive practices and build visitors' agency, educators and designers of informal learning environments might consider the following questions:

- In exhibits, programs, or other informal experiences, when and how can learners make choices? What parts of the process can learners contribute to?
- Are learners' choices consequential? Is the process or the outcome predetermined, or is it shaped by learners' questions and ideas?
- When and how is scientific information presented? Is it a starting point or an end point of the experience?

Who Am I? Agency as a Transformative Experience

The research described above suggests that environments can be designed to support agency by prompting learners to take action, make choices, and engage in conversations. However, learners' internal perceptions of their involvement are equally important. Learners' sense of agency depends not only on the roles available to them in a given setting but also on the roles they choose to take on (Bandura, 2006; Miller et al., 2018). Researchers have described this expression of agency as an ongoing part of identity development. The choices learners make about how to participate in a learning activity or environment allow them to explore and develop parts of their identities and to stretch their boundaries by experimenting with new roles and ways of being (Barton & Tan, 2010).

Connecting to Visitors' Identities

In museum settings, visitors foreground slightly different aspects of their identities depending on their motivations for each visit and the social and physical contexts they encounter (Falk, 2006). In these settings, learners exercise agency by choosing what to explore based on their existing interests and by making sense of

susan m. letourneau

their reactions to unexpected experiences that spark their curiosity in the moment (Rounds, 2006). Although one visit to a museum may not be transformative in itself, it is part of an ongoing transformative process in which learners simultaneously reinforce and expand their perspectives about who they are and who they could be (Rounds, 2006). In science education, this kind of ongoing transformation is critical to developing visitors' emerging interests and identities. Each learning activity provides opportunities for learners to understand science as a community of practice and see a place for themselves in it (Nasir & Hand, 2006). By actively participating and defining roles for themselves within and across settings, learners test out whether they are the kind of person who might be interested in STEM and whether there is room for their perspectives in STEM fields (Barton & Tan, 2010; Gutiérrez, 2008).

Chapter 3 describes a design-based research project at NYSCI to address concerns that girls opt out of pursuing engineering because they perceive it to be impersonal, competitive, and disconnected from their interests and values. Activity developers and researchers worked together to reframe engineering design activities to place empathy for others at the center. This created more gender-inclusive experiences that focused on helping others rather than solving physical or technological challenges. Presenting empathy as an integral part of the engineering design process was effective in inviting girls to tackle design challenges and use engineering practices. This project created space for learners across genders to leverage socioemotional skills that are often neglected in science education but are nonetheless critical for success in many scientific fields. Ultimately, these kinds of experiences may shift learners' perspectives about what STEM learning entails and about themselves as STEM learners.

Together, this research suggests that learning environments can support agency not only by providing a range of experiences to choose from but also by inviting learners to express aspects of who they are and offering approachable opportunities for them to participate in less familiar ways without judgment or pressure. This work also points to the need to present STEM as a field that welcomes, and in fact requires, a wide range of perspectives, approaches, and ways of knowing. There are several ways to open up learners' perspectives about whose ideas are welcome and what roles are possible. Examples include immersive experiences that allow learners to see themselves and the world around them from a new point of view; opportunities to use real scientific tools, be creative in ways one might never have attempted, or learn from and with others; and representation of people from diverse backgrounds in STEM programs and exhibitions.

Empowering Personal Choice

This line of research also implies that learners can exercise agency by choosing *not* to engage or by participating in ways that subvert the intentions of educators, designers, and activity developers (Barton & Tan, 2010; Rounds, 2006). In museums, visitors frequently use exhibits in ways they were not designed for. Visitors may ignore instructions or guidance or bounce from one exhibit to the next without spending much time on any one experience. Groups of students visiting museums on field trips, for example, may prioritize interactions with their peers, imaginative story-telling, and large-scale kinesthetic experiences, which at times may be contrary to the agendas of teachers and museum educators (Anderson et al., 2002; Davidson et al., 2010). Learners also respond to aspects of exhibits in very idiosyncratic ways that are difficult for designers to predict (Falk, 2006). For example, adult visitors may want to revisit exhibits that they remember from their youth and share memories with their own children. These memories may have little to do with the learning goals that designers expected but are nonetheless supporting families' shared histories and identities.

The work described above suggests that asserting one's point of view and pushing back against the expectations of those "in charge" is an intentional and very personal choice. The tension here lies in the need to use STEM as a context for both reinforcing existing aspects of learners' identities and expanding learners' perceptions of themselves and the world around them. One proposed avenue for accomplishing this is to make connections to learners' own goals for themselves and their communities; in this approach, STEM is presented as a tool for expressing oneself and taking social action to create change (Turner & Font, 2007; Barton & Tan, 2010). For example, in "citizen science" and community action initiatives, members of the public gather data on scientific topics that relate to larger societal issues like climate change, public health, and local ecosystems (McCallie et al., 2009). By cultivating learners' agency in addressing scientific issues that are relevant to their lives, these types of experiences can show learners that their actions and ideas are consequential – setting the stage for continued engagement and action.

One such example is NYSCI's Innovation Institute, described in Chapter 8. This program fostered high school students' STEM identities by engaging them in using design thinking and computational tools to create social change. The program helped students build technical skills, but more importantly, it encouraged them to base their projects on their own observations about the assets and needs within their communities. By presenting STEM as a tool that students could use to address issues they cared about, the program fostered their agency in using scientific knowledge to create change, as well as their understanding of the options available to them if they chose to pursue STEM careers.

Questions for Practice

The following questions may be helpful for practitioners that seek to emphasize agency as a transformative experience:

- What opportunities do learners have to connect to and express different facets of their identities?
- How do learners choose to participate? What roles do they take on, including and beyond what was expected or intended?
- What connections could be made to learners' interests, values, or communities?

Agency in Family Groups

Issues of agency and inclusion are more complicated when the target audience consists of multigenerational groups and diverse family structures. Research suggests that caregivers often visit museums for their children rather than themselves (Wilkening, 2009). Nevertheless, families typically function as systems: caregivers' engagement affects children's learning and vice versa (Ellenbogen et al., 2004). A great deal of research has shown how families learn together in museums by building on their shared experiences and cultural practices (Ash, 2004; Crowley et al., 2001; Rogoff et al., 2016). Likewise, agency in STEM learning has been described as a fundamentally social practice driven by conversations and actions within groups of learners as they make sense of new ideas and solve problems together (Arnold & Clarke, 2014; Damsa et al., 2010; Miller et al., 2018; Stroupe, 2014). When families visit museums, then, do they have control over how they learn together?

Considering Caregivers' Needs, Goals, and Assumptions

Practitioners in museums and science centers may expect caregivers to guide their children's learning or allow children to take the lead, or both (Gaskins, 2008). Yet, studies suggest that caregivers' needs and goals for their family experiences vary along many dimensions and do not always align with museums' expectations (Downey et al., 2010; Gaskins, 2008; Letourneau et al., 2017; Wood & Wolf, 2010). For example, caregivers frequently choose to stand back and let children explore on their own (Wood & Wolf, 2010). In spaces designed for young children, adults may not feel comfortable playing with their children in front of others (Downey et al., 2010), or they may prioritize children's independence or social interactions with their peers and intentionally minimize their own involvement (Letourneau et al., 2017). Caregivers take on a wide range of goals in museum settings – teaching, collaborating, scaffolding, playing, observing, and many others (Beaumont, 2010; Gaskins, 2008; Swartz & Crowley, 2004). More generally, the ways in which caregivers are involved in children's learning vary considerably across cultures, as do families' cultural practices and ideas about the nature, origins, and limits of scientific knowledge (Medin & Gang, 2014). This can make it challenging for many caregivers to leverage their own knowledge and experience to support their children's learning in informal spaces (Ash & Lombana, 2011; Gaskins, 2008; Medin & Bang, 2014).

Even though practitioners' assumptions about how families will interact may not be intentional or explicit, these assumptions can become embedded in the design or facilitation of informal learning environments (Allen & Gutwill, 2016; Rogoff et al., 2016). This can affect families' feelings about whether they "belong" in these spaces and the choices they are able to make about how they learn together. Indeed, museums' implicit expectations are not lost on families – studies suggest that caregivers may feel unsure about how they should be involved in children's learning in museums (Downey et al., 2010). At times, they may reject the support offered by facilitators if

Figure 1.2 A caregiver helps her child make sense of the collectively gathered measurements of exhibits to decide which exhibit is the smallest or biggest.

they see it as intrusive or overly didactic (Pattison & Dierking, 2012). Empowering families to exercise agency in how they structure their interactions with one another is critical to enabling them to pursue new ideas and experiences in ways that are meaningful for everyone in their group and that build on their shared experiences in other settings.

For example, NYSCI's Big Data for Little Kids program, described in Chapter 6, aimed to introduce children and their caregivers to concepts about data through activities in which families gathered information about museum exhibits. Rather than telling caregivers how they should be involved, facilitators gave caregivers information about the goals and purpose of each activity, as well as options for how they could approach it as a family. This strategy encouraged caregivers to be co-facilitators throughout the program (see Figure 1.2). Caregivers were able to choose how to adjust the activities to suit their families' needs and make connections to children's existing interests and ways of learning.

Questions for Practice

Several questions may help practitioners as they consider how to strengthen agency for family groups:

- Are experiences for family audiences designed to support both children and their caregivers?

- What roles are available for caregivers to take on? What expectations or assumptions do practitioners have about how families will learn together or how adults will be involved?
- What opportunities exist for caregivers and children to build on their shared experiences and for families to use familiar cultural practices?

What Does Agency Look Like?

As the research summarized here shows, striving to support learners' agency is a complex and nuanced process – one that is deeply connected to issues of equity and inclusion. Although there are many ways of thinking about agency in informal learning environments, agency is often recognized through the qualities of learners' interactions with one another and the surrounding environment. Researchers in science education have described agency as a social practice that is enacted through conversations and actions within a learning environment. These interactions can reveal learners' intentions, motivations, and reactions to the goals and expectations of educators or designers (Arnold & Clarke, 2014; Damsa et al., 2010). Certain interactive aspects of agency are visible in informal settings, enabling researchers and practitioners to investigate what practices might better support them. Some of these aspects include the following:

- The choices learners make about what to explore, what's important to know, or what questions to ask (Bandura, 2006; Miller et al., 2018)
- How learners work together to make sense of new ideas or solve problems (Arnold & Clarke, 2014; Damsa et al., 2010)
- How learners contribute to a shared learning activity, and whether their contributions shape the direction or outcome of the activity in meaningful ways (Barton & Tan, 2010; Miller et al., 2018; Stroupe et al., 2014)
- The roles learners choose to take on and their perceptions of their own involvement, interests, and capacities (Barton & Tan, 2010; Miller et al., 2018)
- The connections learners make between STEM concepts and practices and their own identities and cultural practices (Medin & Bang, 2014; Nasir et al., 2006; Vossoughi et al., 2013).

Conclusion

Taken together, these lines of research offer a theoretical lens that can reframe and deepen our understanding of how informal environments might support STEM learning. This research does not suggest there is a single correct approach. Rather, it prompts researchers and practitioners to question current practices and explore new possibilities for learners to make meaningful contributions to the learning process. Developing engaging and inclusive practices that allow visitors to exercise greater agency over what and how they learn requires critically examining what agency means in informal spaces, and which aspects of informal learning experiences might support or limit it and for whom.

The findings and examples in this chapter provide starting points for considering the topic of agency from different points of view in order to guide further research and practice. Reframing science center experiences to engage visitors in using STEM practices on their own terms may mean that visitors ask questions or reach answers that we did not anticipate or design for. But this work illustrates the many decision points available to practitioners to open up new opportunities for deeper learning.

References

Aikenhead, G. S. (2006). *Science education for everyday life: Evidence-based practice.* Teachers College Press.

Allen, S. (2004). Designs for learning: Studying science museum exhibits that do more than entertain. *Science Education, 88*(S1), S17–S33. https://doi.org/10.1002/sce.20016

Allen, S., & Gutwill, J. (2016). Exploring models of research-practice partnership within a single institution: Two kinds of jointly negotiated research. In D. M. Sobel & J. L. Jipson (Eds.), *Cognitive development in museum settings: Relating research and practice* (pp. 190–208). Psychology Press.

Anderson, D., Piscitelli, B., Weier, K., Everett, M., & Tayler, C. (2002). Children's museum experiences: Identifying powerful mediators of learning. *Curator: The Museum Journal, 45*(3), 213–231.

Arnold, J., & Clarke, D. J. (2014). What is "agency"? Perspectives in science education research. *International Journal of Science Education, 36*(5), 735–754. https://doi.org/10.1080/09500693.2013.825066

Ash, D. (2004). Reflective scientific sense-making dialogue in two languages: The science in the dialogue and the dialogue in the science. *Science Education, 88*(6), 855–884. https://doi.org/10.1002/sce.20002

Ash, D., & Lombana, J. (2011, April 5). *Reculturing museums: Scaffolding towards equitable mediation in informal settings* [Paper presentation]. National Association for Research in Science Teaching Equity and Ethics Symposium, Orlando, FL.

Bandura, A. (2006). Toward a psychology of human agency. *Perspectives on Psychological Science, 1*(2), 164–180.

Barton, A. C., & Tan, E. (2010). We be burnin'! Agency, identity, and science learning. *The Journal of the Learning Sciences, 19*(2), 187–229. https://doi.org/10.1080/10508400903530044

Beaumont, L. (2010). *Developing the adult child interaction inventory: A methodological study.* Center for Advancement of Informal Science. www.informalscience.org/sites/default/files/Preschoolers_Parents_and_Educators-Developing_the_Adult_Child_Interaction_Inventory.pdf

Buechley, L. (2013, October 27–28). *Thinking about making* [Conference presentation]. FabLearn Conference 2013, Stanford University, Palo Alto, CA. http://edstream.stanford.edu/Video/Play/883b61dd951d4d3f90abeec65eead2911d#

Crowley, K., Callanan, M. A., Jipson, J. L., Galco, J., Topping, K., & Shrager, J. (2001). Shared scientific thinking in everyday parent-child activity. *Science Education*, *85*(6), 712–732.

Damsa, C. I., Kirschner, P. A., Andriessen, J. E. B., Erkens, G., & Sins, P. H. M. (2010). Shared epistemic agency: An empirical study of an emergent construct. *Journal of the Learning Sciences* (Vol. 19). https://doi.org/10.1080/10508401003708381

Davidson, S. K., Passmore, C., & Anderson, D. (2010). Learning on zoo field trips: The interaction of the agendas and practices of students, teachers, and zoo educators. *Science Education*, *94*(1), 122–141. https://doi.org/10.1002/sce.20356

Dawson, E. (2014). "Not designed for us": How science museums and science centers socially exclude low-income, minority ethnic groups. *Science Education*, *98*(6), 981–1008. https://doi.org/10.1002/sce.21133

Dawson, E. (2017). Social justice and out-of-school science learning: Exploring equity in science television, science clubs and maker spaces. *Science Education*, *101*(4), 539–547. https://doi.org/10.1002/sce.21288

Dawson, E. (2019). *Equity, exclusion, and everyday science learning: The experiences of minoritized groups*. Routledge Research in Education.

Dilenschneider, C. (2017, April 26). People trust museums more than newspapers. Here is why that matters right now. *Know your own bone*. www.colleendilen.com/2017/04/26/people-trust-museums-more-than-newspapers-here-is-why-that-matters-right-now-data/

Downey, S., Krantz, A., & Skidmore, E. (2010). The parental role in children's museums. *Museums and Social Issues*, *5*(1), 15–34. http://doi.org/10.1179/msi.2010.5.1.15

Ellenbogen, K. M., Luke, J. J., & Dierking, L. D. (2004). Family learning research in museums: An emerging disciplinary matrix? *Science Education*, *88*(S1), S48–S58. https://doi.org/10.1002/sce.20015

Falk, J. H. (2006). An identity-centered approach to understanding museum learning. *Curator: The Museum Journal*, *49*(2), 151–166.

Falk, J. H., & Dierking, L. D. (2016). *The museum experience revisited*. Routledge.

Gaskins, S. (2008, Spring). Designing exhibitions to support families' cultural understandings. *Exhibitionist*, 11–19.

Gutiérrez, K. D. (2008). Developing a sociocritical literacy in the third space. *Reading Research Quarterly*, *43*(2), 148–164.

Gutiérrez, K. D., Schwartz, L., DiGiacomo, D., & Vossoughi, S. (2014, April 3–7). *Making and tinkering: Creativity, imagination, and ingenuity as a fundamental human practice* [Paper presentation]. American Educational Research Association 14th annual meeting, Philadelphia, PA.

Hein, G. (1998). *Learning in the museum*. Routledge.

Hein, G. (2006). Museum education. In S. Macdonald (Ed.), *A companion to museum studies* (pp. 340–352). Blackwell Publishing Ltd.

Humphrey, T., & Gutwill, J. (2005). *Fostering active prolonged engagement: The art of creating APE exhibits*. Left Coast Press.

Letourneau, S. M., Meisner, R., Neuwirth, J., & Sobel, D. M. (2017). What do caregivers notice and value about how children learn through play in a children's museum? *Journal of Museum Education, 42*(1), 87–98. https://doi.org/10.1080/10598650.2016.1260436

McCallie, E., Bell, L., Lohwater, T., Falk, J. H., Lehr, J. L., Lewenstein, B. V., Needham, C., & Wiehe, B. (2009). Many experts, many audiences: Public engagement with science and informal science education. A CAISE inquiry group report. Center for Advancement of Informal Science Education. www.informalscience.org/sites/default/files/PublicEngagementwithScience.pdf

McLean, K., & Pollock, W. (Eds.). (2007). *Visitor voices in museum exhibitions.* Association of Science-Technology Centers Incorporated.

Medin, D. L., & Bang, M. (2014). *Who's asking? Native science, Western science, and science education.* MIT Press.

Miller, E., Manz, E., Russ, R., Stroupe, D., & Berland, L. (2018). Addressing the epistemic elephant in the room: Epistemic agency and the Next Generation Science Standards. *Journal of Research in Science Teaching, 55*(7), 1053–1075. https://doi.org/10.1002/tea.21459

Nasir, N. S., & Hand, V. M. (2006). Exploring sociocultural perspectives on race, culture, and learning. *Review of Educational Research, 76*(4), 449–475.

Nasir, N. S., Warren, B., Rosebery, A., & Lee, C. (2006). Learning as a cultural process: Achieving equity through diversity. In K. Sawyer (Ed.), *Cambridge handbook of the learning sciences* (pp. 489–504). Cambridge University Press.

National Academies of Sciences, Engineering, and Medicine. (2018). *How people learn II: Learners, contexts, and cultures.* The National Academies Press. https://doi.org/10.17226/24783

Pattison, S. A., & Dierking, L. D. (2012). Exploring staff facilitation that supports family learning. *The Journal of Museum Education, 37*(3), 69–80. https://doi.org/10.1080/10598650.2012.11510743

Penuel, W. R., & Furtak, E. M. (2019). Science-as-practice and the status of knowledge: A response to Osborne. *Science Education, 103*(5), 1301–1305.

Rogoff, B., Callanan, M., Gutiérrez, K. D., & Erickson, F. (2016). The organization of informal learning. *Review of Research in Education, 40*(1), 356–401. https://doi.org/10.3102/0091732X16680994

Rounds, J. (2006). Doing identity work in museums. *Curator: The Museum Journal, 49*(2), 133–150.

Russ, R. S., & Berland, L. K. (2019). Invented science: A framework for discussing a persistent problem of practice. *Journal of the Learning Sciences, 28*(3), 279–301.

Sheridan, K., Halverson, E. R., Litts, B., Brahms, L., Jacobs-Priebe, L., & Owens, T. (2014). Comparative case study of three makerspaces. *Harvard Educational Review, 84*(4), 505–532. https://doi.org/10.17763/haer.84.4.brr34733723j648u

Simon, N. (2010). *The participatory museum.* Museum 2.0. www.participatorymuseum.org/read/

Stroupe, D. (2014). Examining classroom science practice communities: How teachers and students negotiate epistemic agency and learn science-as-practice. *Science Education, 98*(3), 487–516. https://doi.org/10.1002/sce.21112

Stroupe, D., Caballero, M. D., & White, P. (2018). Fostering students' epistemic agency through the co-configuration of moth research. *Science Education, 102*(6), 1176–1200. https://doi.org/10.1002/sce.21469

Swartz, M. I., & Crowley, K. (2004). Parent beliefs about teaching and learning in a children's museum. *Visitor Studies Today, 7*(2), 1–16.

Turner, E. E., & Font, B. T. (2007). Problem posing that makes a difference: Students posing and investigating mathematical problems related to overcrowding at their school. *Teaching Children Mathematics, 13*(9), 457–463.

Vossoughi, S., Escudé, M., Kong F., & Hooper, P. (2013, October 27–28). *Tinkering, learning & equity in the afterschool setting* [Paper presentation]. FabLearn conference 2013, Stanford University, Palo Alto, CA.

Whitcomb, A. (2006). Interactivity: Thinking beyond. In S. Macdonald (Ed.), *A companion to museum studies* (pp. 353–361). Blackwell Publishing Ltd.

Wilkening, S. (2009, January–February). Moms, museums, and motivations: Cultivating an audience of museum advocates. *ASTC Dimensions.* www.astc.org/astc-dimensions/moms-museums-and-motivations-cultivating-an-audience-of-museum-advocates/

Wood, E., & Wolf, B. (2010). When parents stand back is family learning still possible? *Museums and Social Issues, 5*(1), 35–50. https://doi.org/10.1179/msi.2010.5.1.35

Zimmerman, H. T., Reeve, S., & Bell, P. (2010). Family sense-making practices in science center conversations. *Science Education, 94*(3), 478–505. https://doi.org/10.1002/sce.20374

chapter 2

From Explaining to Engaging Visitors
Transforming the Facilitator's Role

*Priya Mohabir, Dorothy Bennett,
C. James Liu, and Daisy Tetecatl*

> The [training] program shifted to focus more on the professional development aspect in building a skill set to interact with a variety of audiences. And that, I think, is the most visible example of how our training has changed. Back when I first started, our training was much more content-focused. We would spend one week just strictly learning relevant content about a certain exhibit area, and the following week, we would practice explaining our understanding to each other. We started gravitating away from an exclusive focus on science content to methods that involved engaging our visitors. The training was reformatted to focus more on the ways in which you can start talking to someone about an exhibit.
>
> A museum "Explainer"

Over the last 10 years, the New York Hall of Science (NYSCI) has been reimagining our museum visitor experience. We have moved away from developing stand-alone exhibits and outreach programs focused on fostering understanding of science, technology, engineering, and mathematics (STEM) concepts and phenomena toward more playful, participatory experiences. Visitors are encouraged to explore, build skills, design and make things, and solve problems. Guided by research in developmental and cognitive psychology and the learning sciences (National Academies of Science, Engineering, and Medicine [NASEM], 2018; National Research Council, 2000), we have emphasized engagement that draws out visitors' personal interests and encourages creative problem-solving.

To bring about this change, we have gradually transformed the roles of our young museum floor staff, known as Explainers. In practice, Explainers have shifted their focus from explaining the STEM content and concepts behind exhibits to supporting visitors in following their own questions and designing their own solutions. This has also affected personnel in museum departments that rely on and support Explainers, including exhibit developers, youth developers and trainers, researchers, and public program staff. These departments needed to collectively identify and develop a framework of the skills and mindsets that could support and promote Explainers' own agency and would result in more inclusive and engaging STEM experiences for visitors.

In this chapter, we discuss NYSCI's process for revamping the Explainer role, as well as some of the strategies and challenges involved in rethinking what it means to actively engage visitors with STEM. We share insights from an ongoing multi-year museum initiative to support museum facilitators in making this shift. We draw on formative research conducted with our Explainers about their expectations for and conceptions of their roles and practices at different points during this evolutionary process. We also share information from documentation and interviews with training staff about the design and implementation of new forms of training, different ways of identifying what Explainers should know and be able to do, and new forms of cross-departmental collaboration. We end with implications for the museum field about how to refine practices to support more visitor agency and community-based interaction with STEM while still incorporating STEM content and big ideas.

The Beginning

In 2010, NYSCI embarked on a journey toward more visitor-centered STEM engagement when it became an early partner in hosting World Maker Faire. This family-friendly event is "part science fair, part county fair . . . where tech enthusiasts, crafters, educators, tinkerers, hobbyists, engineers, science clubs, authors, artists, students, and commercial exhibitors [gather] to show what they have made and to share what they have learned" (Maker Faire, 2020). At that time, the maker movement was gaining traction and was helping to promote a do-it-yourself mindset. An underlying premise is that innovation and authentic learning start with deep engagement and exploration and that people learn as they design and make things for themselves and others.

Building on this momentum, our museum set out to create new, imaginative exhibit spaces to support learning through self-directed exploration, design, and making – spaces that would nurture the joy of learning that is intrinsic in young people's play. In essence, we wanted our spaces to foster visitors' agency. We wanted to provide access to materials, tools, and resources for visitors to work on problems they find worth solving and discover their inner inventiveness, creativity, and inquisitiveness. We were especially interested in reaching reluctant learners who don't think of themselves as people who are good at STEM. Dubbed Design Make Play, this learner-centered pedagogical approach reflected findings from the science of learning about essential elements for deeper learning and core ideas for visitor engagement. (See this book's Introduction for more details about the Design Make Play principles.)

Our new approach to pedagogy guided the development of two large exhibit spaces in 2014–2015: a 5,250-square-foot Design Lab dedicated to creative problem-solving with common materials in novel ways; and Connected Worlds, a digitally immersive environment where visitors work together to try to grow and sustain ecosystems and keep a simulated world in balance. Not surprisingly, with these new spaces came new challenges. In keeping with NYSCI's origins as a Hall of Science at the 1964 World's Fair, a large proportion of our exhibits have long focused on getting visitors to explore scientific phenomena and eliciting "aha" moments about the big ideas in science. The transition to collaborative design, tinkering, and open-ended discovery that is defined largely by the visitor's own interests was unsettling for some visitors and staff. This was especially true for our Explainers and other museum staff who regularly helped Explainers learn how to communicate and build on the content of our exhibits. In light of these new exhibitions, we needed to rethink what facilitation on the floor looked like.

When we started this journey, there was limited research on how to prepare museum facilitators to think differently and adapt to new roles; everyone was still wedded to a "right answer" paradigm. This process required altering the facilitator's role from acting as knowledgeable science expert to interacting with visitors in ways that foster their agency and STEM literacy. For NYSCI, this meant empowering Explainers to take ownership of our Design Make Play guiding principles and see the value of putting visitors at the center of their own learning – goals that even the best science teachers struggle to achieve.

Evolving from Helping Visitors "Get" the Science to Getting Them to Explore Science

Explainers are participants in the Science Career Ladder (SCL), a signature NYSCI program that combines youth development and youth employment. Founded in 1986, SCL has been a model for at least two dozen similar programs worldwide. During a typical year of this program, up to 150 high school and college students are employed as Explainers on the museum floor. They advance through a system of job opportunities, scaffolded by a robust program of ongoing professional development and STEM learning opportunities. They receive college readiness workshops, exposure to STEM careers and academic opportunities, and mentorship from program alumni, near-peers, and museum professionals.

Explainers are a valued and highly visible part of the NYSCI experience. They help a diverse community of visitors learn science skills and concepts by conducting captivating demonstrations, facilitating exhibits, and engaging visitors in fun, hands-on activities that ignite a passion for STEM. The Explainers themselves represent a diverse

priya mohabir et al.

community of learners – 85% are from racial-ethnic backgrounds underrepresented in STEM fields, and 65% identify as female and 35% as male. About 47% are enrolled in high school and 53% are in college. Most of them (87%) are from Queens, one of the most ethnically diverse counties in the United States.

The SCL and NYSCI came of age together and have continued to evolve over time. Early on, NYSCI recognized that "explaining" was really about teaching science. To fulfill the call for high-quality science educators at the time of SCL's inception (AAAS, 1989), we supported many SCL participants in career pathways that would lead to their becoming part of a diverse new generation of science educators who were well-equipped to teach complex concepts and skills. Largely designed by museum education staff, as well as staff members who were former Explainers themselves, our Explainer training programs subsequently focused on helping young facilitators understand the science behind our nearly 450 stand-alone exhibits, similar to those found in science centers across the country. Playful and potentially awe-inspiring, many of these exhibits were purposefully designed to allow visitors to experience and discover a principle or fact about a scientific phenomenon, such as how the eye sees, how light refraction makes rainbows, or how sound travels. Explainers interacted with visitors by following patterns used by many informal science educators – for example, by opening with questions to spark curiosity, providing information that explained what was happening, and connecting a phenomenon to something the visitor might be familiar with.

In this context, the training aimed to build Explainers' knowledge of the science concepts embodied in the exhibits and to teach them effective strategies for relating these concepts to visitors' everyday lives. The ultimate goal was science communication, and the training provided at the time reflected this.

It became quickly apparent, however, that this old training paradigm did not match our new exhibits like Design Lab, where visitors are invited to engage in engineering practices to design solutions to problems they define themselves. Although Explainer training included strategies for engaging different audiences and age groups, it focused far more on the content we wanted visitors to learn. We realized that in order to facilitate the kinds of visitor agency we aspired to, Explainers would need to employ a different set of skills. They would need to step back and allow visitors to take the lead. They would need to develop more nuanced and fluid strategies to further visitors' thinking and assess when to step in and when not to.

Promoting Explainer Agency

To prepare Explainers to embrace Design Make Play in practice, we focused on resetting expectations about their role, expanding support networks, and elevating their voices. We quickly realized that we needed to start with where Explainers were while enlisting Explainers as active collaborators and contributors to this new mission and purpose. This process of resetting expectations for Explainers would take time and would depend on their acceptance of a different goal for visitor engagement. It also required an institutional effort to re-envision how to share authority with Explainers.

A critical starting point for promoting Explainers' agency was to rethink the training sessions that the SCL team provided. The emphasis of Explainer training had to shift from knowing the content to providing opportunities for visitors to engage with science and engineering *practices*. This required a series of structural and organizational changes as well as changes to well-established pedagogical practices for preparing our young facilitators.

Starting in Design Lab and Including Perspectives of Senior Explainers

With Design Lab, we started by working collaboratively across our exhibit, research, public program, and youth development departments. We set out to unpack what a Design Make Play approach to STEM engagement would mean in our museum and, more importantly, determine how to prepare Explainers to put this approach into action. Members from each department met regularly to identify engineering design practices that would engage visitors in creative problem-solving and highlight the big ideas rather than explaining scientific principles. Design Lab offered the perfect laboratory for this transformation. We had to consider how to realize a very different end goal from anything we had done before in the museum. We also pushed our colleagues to articulate what was most important for Explainers to know and be able to do in Design Lab and how this related to other experiences throughout the museum.

As members from these departments continued to meet to develop strategies for supporting Explainers in understanding their changing roles, we began to expand the role of Senior Explainers. These are individuals in the third tier of the SCL who have been in the program an average of three to four years. The trainers and the Design Lab staff recognized early on that the perspectives of Senior Explainers would be critical in shaping future training experiences. The Senior Explainers could help translate our new goals for visitor engagement into real practices on the museum floor because they understood intimately what this shift required and what kinds of questions Explainers were likely to have. As a result, Senior Explainers became the backbone of support in training, offering feedback from their own experiences interacting with visitors. Eventually the trainers began to step back as the Senior Explainers led small peer-to-peer coaching sessions, called "floor prep," to help newer Explainers gain real practice and obtain constructive feedback about their interactions with visitors.

Updating Orientation for Explainers

While Explainer orientation always aimed to set a playful tone for visitor engagement and to introduce new Explainers to a supportive community, we realized we had to do more to introduce new Explainers to the learning culture we aimed to cultivate across the museum. We updated our three-day, 15-hour orientation to include opportunities for new Explainers to first experience the museum as a visitor and reflect on their own learning experiences in Design Lab and on the floor. In particular, the museum's youth development staff called out the qualities that made certain things memorable for the new Explainers and observed visitors and veteran Explainers on the museum

priya mohabir et al.

floor. Current Explainers were involved throughout this process of updating orientation – for example, by engaging in reflection about the intentional decisions they make to engage visitors.

From Training to Coaching

Veteran Explainers who were trained on stand-alone exhibits also had to become familiar with a different approach to providing support on the museum floor. Toward this end, we continued to refine training sessions to assist Explainers in facilitating conversation with visitors. In addition, the Design Lab team, many of whom had been Explainers themselves, offered real-time support on the floor. This new opportunity evolved into an informal coaching model and offered Explainers a new lens through which they could notice and reflect on their impacts on visitors.

As explained above, the community of support was expanded to include members of the exhibits department, which was tasked with creating open-ended design activities. These members helped the trainers better understand the intentions of the exhibit space and become learners themselves in this process. While this understanding was largely happening in the Design Lab exhibit, it had a broader impact by helping Explainers better grasp their role and the goals of their interactions throughout the museum.

Shifting from the Content Explainers Need to Know Toward Habits of Mind They Need to Practice

In today's learning environment, where people are often inundated with information, stimulation, and connectivity, we recognized a new sense of urgency to support Explainers in adapting habits of mind that could lead to deeper visitor engagement. These habits of mind include curiosity, engagement, persistence, flexibility, and metacognition. To identify and cultivate these habits, we drew on the work of learning scientists and educational researchers (Costa & Kallick, 2000, 2008; DeBono, 2006; Hmelo-Silver, 2004; Marzano, 1992; NASME, 2018). We also drew on a project at NYSCI, entitled the Formation of Engineers, that explored how to integrate engineering habits of mind into the development and facilitation of engineering design experiences in our Design Lab.

Identifying essential habits of mind made sense for our work with Explainers, who would need to respond nimbly to the diverse needs of visitors with different perspectives, cultural backgrounds, experiences, and interests. We were especially interested in strategies that would enhance ways for visitors to produce their own knowledge rather than merely reproduce information given to them. Explainers would need to develop a critical stance toward their work: inquiring, thinking flexibly, and learning from another person's perspective – not only having information but knowing how to act on it. Identifying habits of mind would also address a long-standing challenge of NYSCI in developing a common language to describe what facilitators do with visitors on the floor. This would be vital for fostering conversations about practice and encouraging Explainers to reflect on their own growth and the growth of their peers.

Drawing on this research, we engaged in an iterative process of refining core values with the team and generating a set of habits of mind that would help build an inclusive museum environment that fosters creativity and innovation. We made the habits of mind descriptive and active so Explainers could envision and name the ways in which they were interacting with visitors. These habits of mind became a sort of mantra, which we made into a list (see Figure 2.1). We distributed the list first to smaller teams of Explainers and then throughout all rungs of the SCL.

ny sci | **Explainer Habits of Mind**

Explainers are **Diverse**.

Our diversity is our greatest asset. It allows us to engage with our visitors from multiple perspectives. We work together to grow and develop the following shared habits of mind to better engage NYSCI's diverse community.

Explainers are **Personable**.

They are friendly, approachable, playful, and open-minded. They nurture an environment that promotes collaboration and sharing ideas.

Explainers are **Empathetic**.

They are good listeners, tuned in to the needs of others and eager to help people explore their unique interests.

Explainers are **Flexible**.

They are flexible decision makers, adjusting their practices to meet the needs of their audience.

Explainers are **Curious**.

They question the world around them, have a desire to learn, and help foster curiosity in others.

Explainers are **Resourceful**.

They use what they have and know in creative ways to meet the situation at hand.

Explainers are **Reflective**.

They reflect on the things that went well, and didn't and feel empowered to change things.

New York Hall of Science

47-01 111th Street
Queens, NY 11368-2950
718 699 0005
Fax: 718 699 1341
www.nysci.org

You're empowered to try things!

Figure 2.1 List of habits of mind distributed to youth employed at the museum.

priya mohabir et al.

To gauge how Explainers might interpret the habits of mind, we convened a group of Explainers to review these habits and gather their thoughts and reactions. Overall, their reactions were positive, but there was some uncertainty about how to develop these behaviors and dispositions. Some of the group felt that Explainers might view the habits of mind as unattainable and assume they would need to be expert in these habits when they entered the SLC program – in other words, they might see the habits as a job description rather than a set of behaviors to be honed over time. Others thought these habits might foster an inclusive museum environment, but they still had questions about how one would hone these skills.

Despite these reservations and questions, the Explainers had recommendations for how to introduce the habits of mind during orientation for new Explainers and ongoing training sessions. Their suggestions included asking Explainers to reflect on their own interactions to identify what habits of mind were at work, setting up training sessions in which Explainers would act out certain habits of mind in mock interactions with visitors, and using the habits of mind as benchmarks of success or guideposts for reflecting on the success of interactions, programs, and events. They felt that introducing the habits just once would be insufficient; instead, these habits should be brought up regularly in assessments and surveys to find out what Explainers were noticing about their own development.

Taking cues from the Explainers, we displayed several iterations of "talk-back" boards in the Explainers' lounge (see Figure 2.2). The boards were intended to prompt Explainers to share examples of what each of the habits looked like in action and how the habits could be translated into interactions with visitors. This yielded some revealing data. Overall, Explainers were very adept at identifying strategies for being personable on the floor, but they still had difficulty integrating empathy and curiosity into their work.

As this process unfolded, the trainers realized that they needed to model what they preached. For example, during orientation sessions for new Explainers, the trainers needed to think about how to be "empathetic," "resourceful," or "curious" in their work with Explainers and call it out during training. The leadership team began to give Explainers more agency during training. Rather than providing everyone with a run-down of their job description and responsibilities, they asked prospective Explainers to write their own job description, share it aloud, and critique it with staff and peers in order to promote reflection. They also gradually altered the training schedule to be much more flexible. In essence, the Explainer leadership team embraced the habits in their own work with Explainers.

In tandem, we began to refine why it was important to make the effort to set new expectations for Explainers. We clarified why Explainers should build their own voice in STEM and allow visitors to do the same, and how training was ultimately a collaborative process with Explainers. Eventually, the habits of mind became guideposts not only for Explainers but for everyone that worked with Explainers on the museum floor. Overall, this transformation alleviated pressure from the trainers to deliver content and encouraged them to include more peer-to-peer learning in their sessions. Orientation sessions were upgraded to allow new Explainers to observe experienced

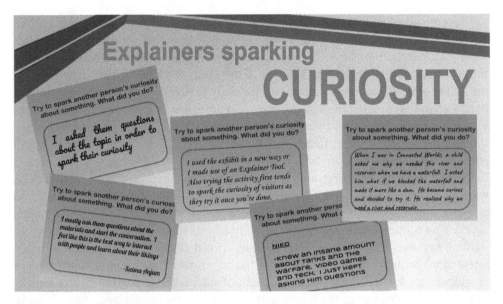

Figure 2.2 Talk-back boards in the Explainer Lounge where Explainers reflect on how they put habits of mind into action on the museum floor.

Explainers as they facilitated interactions. Throughout this process, participants and observers were coached to become comfortable with their discomfort and see critical feedback as a tool for constructive growth.

Interviews with Explainers revealed that these strategies helped to build a supportive environment where Explainers felt more comfortable "not knowing" and asking critical questions. Ultimately, the trainers altered their own interactions with Explainers to emulate the kinds of experiences we wanted Explainers to engage in with visitors. A Senior Explainer said these changes have made Explainer training more "discussion-driven." They helped Explainers become more comfortable with speaking their minds and created a culture in which Explainer training "belongs to the Explainers." She described other benefits of the "very open, peer-led approach" to training:

> It's common practice to get Explainer insight on many projects or general ideas throughout the department, especially with the way a lot of the [Explainer leadership team] members talk to us, too . . . If we have a thought and we say it, they'll usually ask us to elaborate. So it kind of forces us to continue processing information and building our thoughts with the possibility of coming to a solution/conclusion on our own . . . A lot of Explainers gain a sense of responsibility to support and engage each other. For both incoming Explainers and veteran Explainers, they have the opportunity to give and receive feedback not only from the leadership team, but also directly from their peers on all rungs of the ladder.

Key Elements of Our Process

Our formative research suggests that several elements of our transformation process, described below, have been instrumental in promoting agency for Explainers and visitors.

Interdepartmental Collaboration and Institutional Investment in Explainers

In the past, training was led almost exclusively by the trainers and was separated from the work of staff responsible for developing new exhibits, programs, and research and development projects. The collaborative effort between the trainers and the research, public programs, and exhibits teams has been a highly productive strategy for building shared understanding across the organization. It has pushed trainers to think about Explainer preparation in ways they had not before. The biggest outcome of this process has been a more established network of support for developing new training sessions. This includes opportunities to work with the exhibits team to craft approaches that are directly aligned with current exhibit development and to access resources and strategies from research and programs like the game-based facilitation work from the Children's Museum of Pittsburgh.

This collaboration has inspired the Explainer leadership team to provide opportunities for Explainers to assume a more central role in not only shaping facilitation on the museum floor but also designing and refining exhibits and programs to be more visitor-centered and inclusive. One early opportunity involved the museum's public program colleagues, who created a Designer in Residence program. This program paired Explainers with designers, educators, engineers, and artists to co-develop activities, exhibits, and programs for museum visitors. Explainers were more than participants in this program – they were active contributors who brought knowledge and experience in working with our audience.

Shifting Ownership and Expanding Opportunities

The change in Explainer practices from delivering STEM content to listening to visitors and following their lead was new to everyone. It was important to build a culture of experimentation to help Explainers feel supported and empowered to try new things and sometimes fail. Explainers have also been given latitude to choose which activities would go out on the floor each day in our activity-based exhibits and to modify the materials lists to create new opportunities for exploration. They have also contributed to ongoing activity development and research efforts.

As Explainers have become more active contributors to new activities and floor experiences, they have assumed a great deal of ownership of what these experiences are and how they are facilitated. This kind of work has enabled Explainers to see their individuality as an asset they can tap to refine their own facilitation styles and foster engagement that is truly visitor-centered.

Figure 2.3 Explainer following a visitor's lead in our Design Lab's dowels structure activity.

This greater ownership was especially evident in the development of new science demonstrations (see Figure 2.3). Traditionally, Explainers would demonstrate a scientific principle by displaying how something works and following a scripted line of questioning and performative tasks that highlighted the science takeaways, such as demonstrating the properties of matter. In contrast, a new demonstration called Design Time, created collaboratively by the leadership and a team of Explainers, modeled the engineering design process by introducing novel, whimsical materials and inviting visitors to collaboratively design a solution for delivering a ball from one side of the demo space to the other. Ultimately, this experiment inspired a larger reworking of other long-standing, more traditional demonstrations (like cow eye dissections) to give visitors a greater voice and more active role in choosing the direction and outcome of the demonstration. Overall, this work led to positive outcomes that have persisted in the institution.

A Safe Environment in Which Trainers and Explainers Can Experiment

The trainers had to be open to feedback from the cross-departmental members. Much of this work put Explainers and the trainers on equal footing in discussing and exploring what Design Make Play and the habits of mind meant, without a predetermined agenda. The trainers had to change their practices, show a willingness to be vulnerable, take risks, and model behaviors they wanted Explainers to use on and off the museum floor. What made this possible was a sense of trust that we were all in this together and that there were no tried and true approaches. Frequent data from Explainers allowed the trainers to make informed changes to their training practices.

priya mohabir et al.

In addition, other research projects provided data and reflection on our programming. Their contributions helped the trainers increasingly value the role of formative research in refining visitor engagement strategies and offered opportunities to evolve our thinking and practices outside of training. This would not have been possible without support from leadership across the museum departments and a growing understanding of what this meant for our institutional mission to increase visitor agency and inclusion in STEM.

Where Do We Go and What Is the Role of Science Content Now?

At this point, the museum as a whole has been committed to creating training and professional development experiences that prepare Explainers to embrace Design Make Play approaches. These experiences also aim to develop habits of mind that equip Explainers to engage visitors, promote awareness and agency in STEM, and facilitate the learning of big ideas when it makes sense. But young people who come to work in science museums like ours still expect to be responsible for science communication. They don't initially understand that their main role is to be a coach or facilitator and that social skills matter just as much as knowing things. We have come to understand the "why" behind what we are doing, but we have not always clearly shared our constantly evolving understanding with our Explainers. This can lead to changes that do not further the larger institutional mission of promoting agency and inclusion in STEM for all.

Our talk-back boards made clear that Explainers have come a long way with exhibits like Design Lab, which they now see as fun and prefer to facilitate over most other exhibits. But Explainers continue to find our more open-ended playful exploration spaces, such as Connected Worlds, Science Playground, and Preschool Place, to be less satisfying. This suggests that there is more work to do in helping them embrace their role as play and STEM facilitators. Our experience with revamping training points to a number of challenges and future directions for research and development, which are described below.

Using Content as a Tool for Building Connections, Not an End Goal

While we have moved away from direct science communication, many of our exhibits still illustrate science concepts or phenomena. We are realizing that a different kind of "content" preparation is needed for these exhibits. The Next Generation Science Standards (National Research Council, 2013) advocate for learning concepts and big ideas of science in the context of doing science and engineering. Rather than focusing on the facts that Explainers need to know, we need to provide opportunities for Explainers to explore the big ideas behind exhibits and learn how to use science concepts as a tool for making an exhibit relatable and making visitor discoveries more meaningful. Our Explainer team discussed four types of content knowledge that Explainers might need to deepen visitors' engagement, even as they are mindful not to focus on transmitting content as the end goal of the exhibit:

- *Exhibit know-how.* Learning how an exhibit works and becoming proficient in its tools and operation so others can meaningfully interact with it
- *Analogies and real-world examples.* Learning how a visitor's actions and explorations in an exhibit connect to the real world, in order to build bridges from an exhibit to what visitors see every day
- *"Apron" tools.* Learning how a handy set of flexible tools can be used to deepen or extend discoveries that visitors can make within an exhibit and learning when and why to take out these tools
- *Big ideas.* Learning the big ideas behind an exhibit and what the words on exhibit signs actually mean, along with strategies for making these ideas relatable

We envision creating voluntary workshops that highlight tips and tricks for using phenomena-based exhibits and helping visitors get the most out of them. In the bubble area, for example, we encourage Explainers to consider questions such as: How does someone make a very large bubble? What can you help visitors notice? This would also involve curating apron tools that Explainers can use to spark conversation around an exhibit. Finally, we need to explore ways to make connections between the larger concepts of the exhibits and real-world situations that are inclusive and culturally relevant. As a case in point, sports examples may work for some visitors but exclude others. This agenda will include crafting ways of discerning what a visitor likes to do and accessing a broad repertoire of strategies to make connections.

Creating a Facilitation Framework

Many of our museum colleagues have developed facilitation frameworks for their educators and floor staff that identify different phases of visitor engagement and mirror the processes of inquiry, tinkering, or play-based learning (Allen & Crowley, 2017; Gutwill et al., 2015; Wardrip & Brahms, 2015). This has been true for our work in Design Lab, where an engineering process is central. Explainers become familiar with the different parts of the design process so they can help visitors along the way, from identifying a problem they care about to building their own design solutions.

But what about in other areas of the museum? We are working on creating a broader facilitation framework that maps out a process for promoting visitor agency in Design Make Play experiences, while being careful not to have lockstep phases that everyone must go through. We are currently considering a framework that integrates habits of mind and can cross very different exhibit experiences. This framework could then be used to guide Explainers in following the lead of visitors throughout a learning experience by welcoming them, observing, diagnosing, supporting, and reflecting on what visitors want to learn or do.

priya mohabir et al.

Developing a Facilitation Toolkit

We recognized the need for concrete examples of what behaviors and strategies that lead to deeper visitor engagement and agency look like and sound like. We are establishing a small team of more senior Explainers to collect specific examples of different habits of mind in action on the floor and the course of interactions from start to finish in different areas of the museum. We are exploring ways of using social media platforms to collect video and photo exemplars that could spark good discussion and analysis among Explainers in future training sessions.

Implications

Very few Explainers remain in the museum who have experienced the former, more content-driven type of training. The integration of habits of mind and visitor-centered engagement strategies into training has produced encouraging results, according to Explainer surveys and leadership team reports. Explainers' understanding of their flexible and important role on the floor is growing in the intended direction, as illustrated by this comment:

> Explainers are pretty much the face of the museum. [They're the first people] a lot of visitors see when they walk inside the institution, and their primary role is to really engage with visitors, making sure that they have a pleasant time when they're in the museum, and that could come through many different means. So, it could be something as simple as just striking up a conversation with them, assisting them at many of our hands-on exhibits, facilitating workshops or activities in our Design Lab space, or even just engaging with them with any of our science demonstrations.

Another Explainer highlighted additional benefits of the new strategies:

> I think something that's more memorable is when [learners] actually participate in things that they can literally get their hands on, or touch, or play with . . . When people are able to question things, that's like another really important part of learning or being impacted by science. And, I think that's one of the driving forces of science too . . . Those questions are what drive curiosity in people because they are constantly trying to seek out answers to these new questions. Throughout my time here in the SCL I'm much more able to ask those questions . . . I used to be a little bit more reserved but now . . . if something just doesn't make sense to me or something's just fascinating, I have to know why, and I'll go ahead and ask why.

The Explainer leadership team has observed several indicators of success. A real sign of progress has been the growth in Explainers' understanding of our learning goals for visitors. Rather than focusing on visitors leaving with new information, Explainers now emphasize visitors leaving with new questions about STEM, newfound confidence in what they can do, and an awareness of how science and engineering is done.

Explainers no longer seem to ask visitors questions and think that they are done if someone supplies an "acceptable" answer; rather, they are much more inquisitive about visitors' interests and intentions. Trainers have also observed that Explainers are far more personable in their interactions with visitors. Explainers have grown especially skilled in supporting visitors' creativity in Design Lab, as evidenced by the ways in which they provide encouragement, build alongside visitors, and use materials creatively.

Many Explainers also noted that the habits of mind gave them a language and a lens for examining things they do every day on the floor. Some find these habits useful for thinking about their own growth and development, a sentiment echoed by the training team. Trainers report that the habits of mind have helped them put a name to qualities they have always felt were important but assumed people just naturally understood.

Conclusion

Our research work has explored how Explainers perceive themselves as engineers, what they think engineering involves, and who it is for. We need more targeted training that allows Explainers to consider how their actions on the floor communicate what STEM is about and how they are connected to the habits of mind we promote. As the face of the museum, Explainers are critical to our mission to foster visitor agency and inclusion in STEM experiences. Our work as a cross-departmental team has highlighted the importance of providing Explainers with opportunities to exercise their own agency by shaping exhibits, conducting research, and feeling free to experiment and make mistakes – in other words, to engage with STEM as scientists and engineers do. By including Explainers in larger cross-departmental teams, we have moved past seeing Explainers as just part of our youth programs and have come to recognize and value their contributions to our ever-evolving museum.

References

Allen, L. B., & Crowley, K. (2017). From acquisition to inquiry: Supporting informal educators through iterative implementation of practice. In P. Patrick (Ed.), *Preparing informal science educators* (pp. 87–104). Springer.

American Association for the Advancement of Science (AAAS). (1989). *Science for all Americans: Summary, project 2061* (ED309060). ERIC, https://eric.ed.gov/?id=ED309060

Costa, A. L., & Kallick, B. (Eds.) (2000). *Discovering and exploring habits of mind.* Association for Supervision and Curriculum Development.

Costa, A. L., & Kallick, B. (Eds.). (2008). *Learning and leading with habits of mind: 16 essential characteristics for success.* ASCD.

De Bono, E. (2006). *DeBono's thinking course: Powerful tools to transform your thinking.* BBC Active.

Gutwill, J. P., Hido, N., & Sindorf, L. (2015). Research to practice: Observing learning in tinkering activities. *Curator: The Museum Journal, 58*(2), 151–168.

priya mohabir et al.

Hmelo-Silver, C. E. (2004). Problem-based learning: What and how do students learn? *Educational Psychology Review, 16*(3), 235–266.

Maker Faire. (2020, September 23). *A bit of history.* https://makerfaire.com/makerfairehistory/

Marzano, R. J. (1992). *A different kind of classroom: Teaching with dimensions of learning.* Association for Supervision and Curriculum Development.

National Academies of Sciences, Engineering, and Medicine. (2018). *How people learn II: Learners, contexts, and cultures.* The National Academies Press. https://doi.org/10.17226/24783

National Research Council. (2000). *How people learn: Brain, mind, experience, and school.* The National Academies Press.

National Research Council. (2013). *Next Generation Science Standards: For states, by states.* The National Academies Press. https://doi.org/10.17226/18290

Wardrip, P. S., & Brahms, L. (2015, June). Learning practices of making: Developing a framework for design. In M. Umaschi Bers & G. Revelle (Chairs), *IDC '15: Proceedings of the 14th international conference on interaction design and children* (pp. 375–378). Association for Computing Machinery. https://dl.acm.org/doi/proceedings/10.1145/2771839

Narratives, Empathy, and Engineering
Creating Inclusive Engineering Activities

Susan M. Letourneau, Dorothy Bennett, Amelia Merker, Satbir Multani, C. James Liu, Yessenia Argudo, and Dana Schloss

In an activity called Help Grandma at the New York Hall of Science (NYSCI), visitors to the Design Lab build models of novel inventions to solve problems that grandparents face in everyday life, such as carrying groceries or climbing stairs. One seven-year-old participating in this activity designed an invention to help a grandmother named Nonna who kept losing her glasses. After thinking about what would be easy for Nonna to use, she initially sketched a sensor that would seek out the lost glasses. Wanting to make the invention more fun and convenient, she then built a robot with legs that would also bring the glasses back, added ears and a tail to make it resemble a pet, and named her invention the "robot glasses fetcher."

This activity is part of an effort by NYSCI researchers and developers to transform engineering activities in the museum's Design Lab. The intent is to give girls (and children of any gender) approachable starting points and compelling reasons to tackle design problems. Rather than presenting engineering design problems as a series of procedures to be followed or specifications to be met, our approach uses narratives to add human-centered contexts. Learners are invited to think about whom they are designing for and why. Learners take the perspective of a character with a problem, such as the grandmother in the example above, as a way to spark imaginative problem-solving.

In this chapter, we review the research base that informed this work, the collaborative process between researchers and practitioners used to develop and test activities, and the key findings and implications of our work.

Why Focus on Inclusion?

Despite ongoing efforts to make engineering a more inclusive field, women remain persistently underrepresented in engineering professions (Katehi et al., 2009; National Science Foundation, 2019), as do African American and Latinx students and other groups. In part, this is because engineering is typically taught through individual and often competitive engineering challenges that frequently lack any personal or social context (Diekman et al., 2010; Balsamo, 2011; Su & Rounds, 2015). These approaches can limit learners' agency by constraining the problems to be solved and the solutions that are possible and by failing to build on learners' own interests and priorities.

Studies have shown that many groups of learners, including girls, are more likely to express an interest in engineering when it is framed as a collaborative approach to solving problems that are relevant to them or to communities they care about (Bennett, 2000; Ceci & Williams, 2010, 2011; Dorie & Cardella, 2013; Eccles, 2005; Eccles & Wang, 2016; Rusk et al., 2008; Wigfield & Eccles, 2000). Reframing engineering as a profession that involves helping others and tackling socially relevant problems can provide more inviting introductions to the field and can counter perceptions that engineering is impersonal or competitive (Capobianco & Yu, 2014).

This evidence is closely aligned with research and policy shifts in engineering education in general, which have identified socioemotional skills such as empathy as a fundamental and often neglected part of engineering practice (Engineering Accreditation Commission, 2019; Hess & Fila, 2016; Hynes & Swenson, 2013; Walther et al., 2012). This work maintains that solving complex engineering challenges with societal implications requires engineers to empathize with colleagues and clients whose perspectives and needs might differ from their own (Borrego et al., 2013; Capobianco & Yu, 2014, Engineering Accreditation Commission, 2019; Walther et al., 2017).

Empathy is a multifaceted process that includes emotional responses such as feeling compassion and concern, cognitive processes such as considering a situation from another perspective or someone else's point of view, and behaviors intended to benefit others such as taking action to help someone (Baron-Cohen & Wheelwright, 2004; Decety & Jackson, 2004; Preston & DeWaal, 2002). When engineering education emphasizes these aspects of empathy, girls (and children of any gender) can come to see that engineering fundamentally involves working with others to help people, which fosters deeper learning and more positive attitudes toward the field

(Cunningham & Lachapelle, 2016; Lachapelle & Cunningham, 2014; Marra et al., 2016; Terenzini et al., 2001). Even with this breadth of research, we know relatively little about *how* to create engineering activities that evoke empathy and whether this strategy is effective for inviting in girls and other learners who may not be engaged by traditional engineering challenges.

A line of research and development at NYSCI aimed to transform the engineering activities offered in the museum's Design Lab exhibit, an imaginative space for open-ended design and engineering. At 10,000 square feet, it has four thematically distinct areas for different types of engineering design activities. Design Lab welcomes more than 250,000 students annually through schools and other organized groups, in addition to the museum's general audience and family visitors. The activities offered in Design Lab use approachable, repurposed materials and open-ended prompts to engage visitors in defining problems to solve and arriving at divergent solutions. As we observed how visitors approached and engaged with these activities, however, we realized that the purpose of the activities was not always clear. Abstract challenges, such as building an object that can float in a wind tube or designing a device to pick up differently shaped objects, could seem unapproachable or uninteresting to visitors.

We at NYSCI recognized a need to reframe Design Lab activities to give visitors approachable starting points and compelling reasons to tackle design problems. Without this, the activities were unintentionally exclusive, supporting agency only among visitors with pre-existing knowledge and an interest in engineering.

Narrative as a Strategy

Prior research demonstrates that narratives are powerful mechanisms for prompting learners to imagine other situations and points of view (Bruner, 1986). Egan, Bruner, and others argue that narratives evoke empathic responses by clearly communicating a problem or conflict that needs to be solved and the perspectives of the people involved (Bruner, 2002; Egan, 2012; Clark, 2010). Early activity development efforts in NYSCI's Design Lab found that inviting learners to think about whom they were designing for and why helped prompt user-centered thinking and imaginative problem-solving (Bennett & Monahan, 2013). The particulars of the narrative – the small details, specific situation, characters involved, and vivid images – could motivate creative thinking in solving technological design problems.

In addition, this early work showed that complete, structured storylines were not necessary; individual narrative elements, such as a sympathetic character or engaging setting, could shift how visitors approached engineering problems by sparking their imaginations and inviting them to fill in details based on their own ideas and experiences. For example, one early activity in Design Lab asked visitors to use circuits to add something to a model city to make it a happier place (Bennett & Monahan, 2013). This simple prompt provided an inviting context for visitors to imagine the people they were helping with their designs and tell their own stories about the problems they were solving as they engineered structures with cardboard and circuits.

In the next phase of our work, we developed evidence-based strategies for using elements of narratives to evoke users' empathy and support specific engineering practices. In a three-year project, we investigated three interrelated questions:

1. *How can narratives be effectively integrated into engineering activities?* Our goal was to identify strategies for adding narratives to engineering design activities that would support learning of core engineering principles and still leave room for visitors' own ideas and points of view. We wanted to be careful, however, not to limit the appeal of popular engineering activities by adding elaborate or constraining storylines. For example, we wondered how we might create a narrative version of an activity in which visitors built structures with dowels (shown in Figure 3.1). Would it be sufficient to suggest a narrative-based prompt, like asking visitors to build a structure that could protect someone in an earthquake? Would this subtle shift in the framing of the problem affect girls' engagement?

2. *How do narratives evoke different facets of empathy?* We specifically aimed to develop strategies for using narrative elements to evoke the emotional, cognitive, and social facets of empathy. Although narratives can support engagement in other ways – by using whimsy or humor to spur creativity, for example – our

Figure 3.1 Testing out a dowel structure built to withstand an earthquake.

narratives, empathy, and engineering

efforts focused on exploring the theoretical connections between empathy and engineering practices that had been raised in prior research.

3. *What impact do narrative-based activities have on girls' participation and engagement in the engineering design process?* Based on the research described above, we hypothesized that narratives could provide a meaningful context for engaging girls in engineering problems and could support engineering practices by prompting empathy and user-centered thinking.

Research-Practice Collaboration and Activity Development Process

This project integrated knowledge from research and practice in order to balance the theoretical underpinnings of the research with the practical needs on the museum floor. The team included NYSCI researchers, museum staff who develop exhibit activities, and high school and college youth called Explainers who facilitate activities on the museum floor. Each of these groups contributed to decisions throughout the development of the activities. We partnered with three other institutions in order to generate activities and research findings that would be generalizable to the field more broadly. Our partners included the Scott Family Amazeum in Bentonville, Arkansas; the Tech Interactive in San Jose, California; and Creativity Labs at UC Irvine. We brainstormed ideas and shared emerging findings with our partner institutions through yearly in-person project meetings and monthly virtual check-ins.

Our goal was to work together to iteratively develop and test six engineering activities that integrated narratives in different ways to evoke learners' empathy for the users of their designs. We applied design-based research methods and gathered two types of evidence. First, we developed iterative prototypes of each of the six activities; during that process, we documented how narratives could be layered onto engineering design activities and which aspects of the narratives were effective in evoking different facets of empathy. Then, by comparing each narrative activity with a non-narrative version with comparable materials, we investigated the impact of narratives and expressions of empathy on girls' use of engineering practices. The development and testing process for each of the six activities lasted 10 to 12 weeks and included the stages of work described below.

Planning

In weekly meetings of activity developers and researchers, we discussed which activities and classic engineering challenges were promising candidates for incorporating narrative elements. Activity developers at NYSCI and the partner sites offered their perspectives about which existing activities could be developed further and how narratives could be added in "light-touch" ways that made tweaks to existing activities rather than overhauls, and that represented nudges for visitors rather than directives. Researchers provided some parameters for activity development based on the aspects of empathy we wanted to evoke and the engineering practices we hoped to support.

susan m. letourneau et al.

Through rough prototyping with basic materials, we vetted initial ideas and narrowed down which possibilities to test further with visitors. This process required activity developers to become more comfortable with testing activities on a small scale before committing to a design concept. It also required researchers to document rapid changes in activities and incorporate activity developers' perspectives, observations, and priorities as activities took shape. The team negotiated concerns about whether adding narratives to already successful activities might negatively affect visitors' experiences and whether early prototypes and non-narrative versions were sufficiently engaging to be placed on the exhibit floor on a typical busy weekend day. Throughout the process, we wanted to make sure that visitors had a positive experience no matter which version of the activities they happened to engage in.

Prototyping

Activity developers and researchers tested prototype versions of activities with visitors at NYSCI on weekends for four to six weeks. Researchers described the narrative framing for each activity to the Explainers who were facilitating, including the ways in which the activity was designed to evoke empathy, the behaviors or reactions we were looking for from visitors, and the questions the team had about how the activity worked and how visitors might respond. The project team recruited pairs of Explainers to participate in this activity development for several months at a time. This allowed Explainers to see activities evolve and contribute to the development of multiple activities.

As Explainers became more familiar with the purpose of the project, they were empowered to try out new ways of facilitating the activities that could highlight and extend the narratives and could further support empathetic approaches to the engineering design process. Researchers took field notes and interviewed visitors who participated (focusing on girls ages 7–14). At the end of each testing day, researchers, activity developers, and Explainers discussed which strategies worked well and which could be improved for the next week of prototyping.

Testing

Once the basic design of activities was finalized, the team tested the final versions of the activities over an additional six weeks, comparing narrative versions of the activities with versions that lacked narrative elements. Researchers conducted observations and interviews with girls who participated in either the narrative or non-narrative versions. We coded the data to determine whether girls expressed different facets of empathy, including (a) affective responses (expressions of concern or compassion for the user of a design); (b) perspective-taking (imagining what users want or need or how they would use designed solutions); and (c) desire to help (taking action to help the potential user of the design). Table 3.1 gives examples of behaviors demonstrating each of these facets.

We also coded the engineering practices that girls engaged in, including (a) problem scoping (considering multiple constraints or aspects of the problem at

Table 3.1 Indicators of empathy in observational and interview data

Empathy indicator	Examples
Affective responses Talking about the user's feelings; expressing concern, compassion, sympathy	"He looks lonely" while looking at one of the pets in Chain Reaction "That'll make her happy" while thinking about a solution for grandma "I feel bad for him! I hope he's ok!" after testing a design in Safe Landing
Perspective-taking Imagining what users want or need, or how one would use the designed solution	Talking about what it would be like to be in an earthquake while designing a structure Modeling how grandma would use a device to help her open jars
Desire to help Taking action to help the potential user of a design	Reinforcing a dowel structure to protect those inside Making sure grandma is safe or comfortable while using a design

Table 3.2 Engineering practices coded for in observational and interview data

Engineering practice	Examples
Problem scoping Considering multiple constraints or aspects of the overall design problem	Considering both the stability of a dowel structure and making it big enough to fit your whole family Making a design for grandma both easy to use and portable
Ideation Generating multiple possible solutions to the overall design problem	Protecting a falling cell phone using padding, or with a parachute Brainstorming different ways to help grandma carry her groceries
Testing Conducting both large- and small-scale tests of a design's function	Testing one step in a chain reaction, or testing the whole contraption Placing an air-powered vehicle in front of the fan to see how far it goes
Iteration Making changes to a design after a test, either once or repeatedly	Fixing or replacing parts of a chain reaction after testing it Adjusting the sail on an air-powered vehicle after testing it

hand), (b) ideation (generating multiple possible solutions to a problem), (c) testing (conducting a small- or large-scale test of the function of a design), and (d) iteration (changing a design after conducting a test). Table 3.2 shows examples of these practices. Throughout the testing process, we collaboratively refined this coding scheme based on discussions with everyone on the team, identifying indicators of each behavior that we could observe with reliability and that were meaningful to both researchers and practitioners on the team.

susan m. letourneau et al.

Interpreting Data and Generalizing Findings

At the end of each round of activity development, the team reviewed the findings and compared them against what we had observed and attempted in the past. Our goal was to identify and experiment with as many ways of using narratives as possible across the set of six activities. This review process encouraged the team to think creatively about additional strategies for integrating narratives into engineering activities and subtle variations that might make a difference in supporting empathy and engineering practices.

For example, multiple activities that we tested used characters to prompt learners to consider the users of their designs (such as designing something to help a particular character), but different activities used characters in subtly different ways. Some of the characters were realistic (grandparents, pets), while others were whimsical (aliens), and still others were personalized or invented by the children themselves (such as shadow puppets that children created to tell their own stories). By strategically varying how we used these narrative elements from one activity to the next, we enhanced our theoretical understanding of the connections between narratives, empathy, and engineering and developed design principles for effectively integrating narratives into activities. Placing each of the six activities within a larger theoretical framework gave the team a structure and common language for making sense of and applying the findings.

Although this process took longer than anticipated, these cycles of testing and discussion, in which all members of the team were on equal footing, were critical. They helped us to establish strong working relationships in which everyone felt comfortable contributing ideas. The cycles also helped us to understand each other's needs and priorities, such as developers' priority for positive visitor experiences, Explainers' need for support, and researchers' need for data. We have adapted and continued to build on this process in other design-based research projects.

Activity Pairs

The project resulted in a set of six activity pairs that included a version with narrative elements and a non-narrative version with similar goals and materials. The activities integrated narratives in a variety of ways: through characters, settings, problem frames, or some combination of these elements (see Figures 3.1 through 3.4). We also varied the narratives to encourage learners to take different points of view; for example, some activities asked learners to imagine themselves in a novel situation, while others invited them to consider others' needs or problems. Finally, the narratives were communicated in diverse ways, through materials, facilitation, the name of the activity, and other means (Letourneau & Bennett, 2020). The activity pairs included the following:

- *Dowel Structures / Emergency Structures*. Visitors use three-foot dowels and rubber bands to construct a stable structure that can fit everyone in their group (non-narrative) or that can protect their group from an earthquake (narrative). In this case, learners take the point of view of their families in a novel situation. The narrative is conveyed primarily through the problem framing and is communicated by facilitators who provide an introduction to the activity and through materials (signage).

- *Chain Reaction / Help the Pets.* Visitors use a sequence of simple machines to accomplish a goal like ringing a bell or landing a ball in a cup (non-narrative) or to take care of a pet – for example, by feeding or playing with a dog (narrative). Learners take the point of view of a hungry or bored pet. The narrative is expressed through characters and problem framing. It is communicated through materials (realistic models of pets and written prompts about problems learners can solve) and by facilitation that introduces the activity and prompts learners on how to use various simple machines.

- *Invention Challenge / Help Grandma.* Visitors design and build models of novel inventions to solve abstract physical challenges like lifting a heavy object (non-narrative), or solve problems that grandparents face in everyday life, such as carrying groceries or climbing stairs (narrative). Learners take the perspective of a character with a problem, in this case a grandmother. The narrative is communicated by materials in the form of challenge cards or persona cards with details about the user's preferences and needs, by facilitation in the form of verbal instructions, and by other learners through display of their inventions.

- *Dropped Calls / Safe Landing.* Visitors use recycled and repurposed materials to design something to protect a cell phone from a 20-foot drop (non-narrative) or to help an alien or astronaut land safely on a planet (narrative). In both conditions, learners use a cell phone accelerometer app to measure the force of the impact in order to improve their designs. In the narrative version, learners choose and personalize a space character by naming and/or decorating a picture of the character they are designing for. Learners take the perspective of their character, and the narrative is communicated by the materials (character cards and a space-themed backdrop near the testing station) and facilitation (verbal instructions and prompting about the characters' needs and safety).

- *Air-Powered Vehicles / Around the World.* Visitors design a vehicle that uses air to move over different textured surfaces (non-narrative) or to help them travel around the world over different landscapes (narrative). Both conditions have multiple testing stations with different levels of difficulty, from smooth to bumpy. The narrative unfolds in model terrains that resemble tundra, desert, grassland, and forest. Learners take the perspective of themselves in each new setting. The narrative is communicated by the materials in the settings depicted in the testing stations and through verbal facilitation and prompts about what learners might need in each place.

- *Light & Shadow / Shadow Stories.* Using everyday materials, learners create effects with light and shadow (non-narrative) or create shadow puppets and scenes (narrative). This activity allows learners to create their own narratives by building characters and/or settings. The narrative is communicated by the learners themselves through the shadow puppets and backgrounds they create and through the materials of the scenery added by facilitators.

susan m. letourneau et al.

Key Findings

Our final data set included observations and interviews with 190 girls who participated in either the narrative or non-narrative version of these activities. Across all six activities, we found that, overall, girls engaged in more engineering practices in narrative than in non-narrative conditions. We also found that the practices of problem scoping (considering multiple aspects of the problem) and iteration (generating multiple possible solutions) were particularly well-supported in the narrative condition (Letourneau & Bennett, 2020). In addition, an external summative evaluation confirmed that when girls showed at least one indicator of empathy, they stayed longer and demonstrated more engineering practices (Peppler et al., 2020).

Additional qualitative analyses of our observations provided richer detail about how the experience of empathy supported engineering practices. When girls participated in the narrative versions of activities, we observed them alternating between emotionally connecting with the problem and using more analytic problem-solving to work with materials and decide on solutions – a process that research on user-centered design refers to as "mode switching" (Walther et al., 2017). This supported user-centered thinking during problem scoping as girls considered such factors as the user's needs, feelings, comfort, and safety, and during testing and iterating their designs as they imagined how someone would use the design, added features to make designs safer, and made other revisions. For example, in Help Grandma, a seven-year-old participant applied this sort of thinking to design the "robot glasses fetcher" pet described at the beginning of this chapter (shown in Figure 3.2).

When we compared activities, we also found that different narrative elements evoked empathy in different ways. Activities that used characters evoked the strongest expressions of empathy, including emotional and socially beneficial behaviors, such as expressing concern for and a desire to help someone. Activities that emphasized settings evoked empathy less strongly on their own but did prompt perspective-taking, in which visitors imagined what it would be like to be in another place or situation.

By testing and revising the set of six activities, we were able to generate and refine the following design principles for creating evocative and inclusive narrative-based engineering activities:

- *Combine characters and settings to evoke empathy and invite girls in.* Because characters and settings evoked different facets of empathy, they were most effective when combined. We also found that settings were effective in inviting girls in and sparking their interest early in the activity, a hurdle for many original Design Lab activities. For example, in Around the World, the backdrops and landscapes (shown in Figure 3.3) drew visitors in and prompted girls to choose a setting that interested them before thinking about how to design their vehicle.

- *Provide choice in defining users and their problems.* Across all of our activities, we found it was important to use light-touch strategies to anchor a narrative without being overly prescriptive. Instead of using constrained storylines that were limited to a single user or situation, we framed problems with open-ended hints of narrative that allowed children to choose whom to help and what problems to solve. In some activities, such as Help the Pets or Help Grandma, children

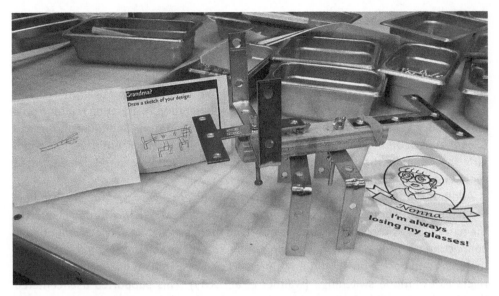

Figure 3.2 A child's creation in Help Grandma: a "robot glasses fetcher" to seek out, grab, and retrieve Nonna's glasses for her.

Figure 3.3 Testing stations in an air-powered vehicles activity, reimagined to use realistic textured terrains including a tundra (left) and a forest (right).

could choose a character to help from a set of options, while in others, such Safe Landing and Shadow Stories, they created characters themselves.

- *Reinforce narratives with materials and facilitation.* We found that children became more immersed in the narratives and engaged in more user-centered thinking when we combined physical reminders of design problems from the users' point of view with prompts from facilitators that referenced and elaborated on the narrative. Both strategies reminded children to think about who would use their design and how their solutions could be more user-friendly. Explainers became adept at using the narrative to guide the prompts and questions at different points in the engineering process. We also found that focusing on imaginative, emotional, and personal aspects of the problem made the process

susan m. letourneau et al.

Figure 3.4 A narrative activity called Help the Pets.

of defining a problem and iterating solutions more approachable for children and caregivers. This was also true for Explainers, who very often struggled to find ways to help visitors who got stuck during the design process. The narrative elements gave Explainers touchpoints for conversations with visitors about whom and what they were designing for and how they could address others' needs. For example, in Help the Pets (shown in Figure 3.4), Explainers invited visitors to name the pets, helped visitors choose materials to incorporate that the pets might enjoy, or talked to visitors about their own pets and what they liked to inspire new ideas.

Implications

The collaborative research and development process used in this project changed our practices and ways of working together. Specifically, we realized that it was critical to foster agency at all levels within our own team in order to open up the activity development process to more voices and perspectives. The project had multiple implications for the activity development process described below.

Supporting Agency Among Visitors and Among Staff

The fact that this project involved consistent collaboration and iteration of ideas over a long period of time influenced our practices in meaningful ways. The repeated activity development process that we engaged in together helped the team think expansively about how existing activities could be transformed, adapted, and reimagined. By working on multiple cycles of rapid activity development, the team devised strategies to test new approaches on a small scale, notice visitors' responses together, and let activities evolve gradually based on the group's shared goals and observations. Making this kind of experimentation a consistent part of our collaborative work pushed us to try things that might not succeed; we brainstormed and tested many different options and began to see failures as an informative part of the process. Although many existing activities were trusted and well-used at the start of the project, the team began to see them as evolving, seeking out opportunities to try new variations and noticing how visitors responded to small tweaks in framing, materials, and facilitation.

These processes of prototyping and testing took place in plain view on the museum floor. Staff and Explainers became accustomed to seeing activities change over time. The Explainers who were directly involved in the project gained agency, as they came to understand what researchers and activity developers were looking for and made active contributions to the creative process. Being a part of this process over time led Explainers to take greater ownership over experimenting with new facilitation approaches elsewhere in the museum and to model this mindset for other Explainers and in other projects. Explainers who were not directly involved in the project also observed how activities in Design Lab changed with input from their peers. The involvement of Explainers as equal members of an interdepartmental team sent a message that their ideas and experiences were valued and necessary to push our thinking further.

Establishing a process in which everyone was on equal footing took a long time but ultimately made our activities more effective and inclusive by incorporating multiple perspectives and areas of expertise. The team embodied the kind of agency and inclusion that we hoped to support in visitors by distributing authority and honoring the unique experiences of each team member. In this sense, supporting visitors' agency meant fostering agency among *staff* at all levels – a lesson that has informed our ongoing work together.

Developing Empathetic and Inclusive Engineering Activities

Our primary goal in Design Lab is to help visitors find design problems they are interested in solving and generate divergent and creative solutions. This project allowed us to test specific and concrete strategies for contextualizing engineering problems and widening their appeal. By adding narrative contexts to the activities in Design Lab, we invited learners to approach engineering by thinking about whom they were designing for and why – an empathetic approach which directly supported user-centered thinking and provided more inclusive entry points into the engineering design process. We specifically focused on girls' engagement in engineering practices, but with the understanding that reframing engineering education in this way would have an impact on many other learners who may be indifferent to traditional engineering challenges that lack a personal or social context.

Considering empathy as an integral part of engineering design fundamentally changed how activities were developed in Design Lab – it became important to think about what visitors were *feeling* while solving engineering challenges, not just what they were doing. We used classic, traditional engineering activities as a foundation, but pushed them further by layering imaginative play, emotional and personal connections, and meaningful real-world situations. Adding narrative contexts provided compelling reasons to tackle design problems and made activities seem as though they had a beginning, a middle, and an end. These realizations shaped the development of Design Lab activities even after the project ended.

Conclusion

This project allowed us to build a shared understanding between researchers and practitioners of how narratives, empathy, and engineering intersect and how these ideas could be implemented and tested in rigorous but feasible ways. We had to work together to understand how different facets of empathy might be expressed in the context of our hands-on engineering activities and what specific engineering practices might be affected. This shared understanding helped us use narratives in strategic and targeted ways that could advance theoretical understanding and simultaneously inform museum practice.

Ultimately this collaborative approach allowed us to develop novel strategies for weaving narratives and empathy throughout the engineering design process. Our work together shows how creating more inclusive invitations into science, technology, engineering, and mathematics and embracing socioemotional skills as an integral part of the engineering process can benefit all learners by showcasing and supporting the full range of skills that scientists and engineers need to solve complex problems with interpersonal, societal, and ethical implications.

References

Balsamo, A. (2011). *Designing culture: The technological imagination at work.* Duke University Press.

Baron-Cohen, S., & Wheelwright, S. (2004). The empathy quotient: An investigation of adults with Asperger syndrome or high functioning autism, and normal sex differences. *Journal of Autism and Developmental Disorders, 34*(2), 163–175.

Bennett, D. (2000). Inviting girls into technology: Developing good educational practices. In *Tech-savvy: Educating girls in the new computer age.* Commission on Technology, Gender, and Teacher Education. American Association of University Women.

Bennett, D. & Monahan, P. (2013). NYSCI Design Lab: No bored kids! In M. Honey & D. Kanter (Eds.), *Design, Make, Play: Growing the next generation of STEM innovators* (pp. 34–49). Routledge.

Borrego, M., Karlin, J., McNair, L., & Beddoes, K. (2013). Team effectiveness theory from industrial and organizational psychology applied to engineering student project teams: A research review. *Journal of Engineering Education, 102*(4), 472–512.

Bruner, J. S. (1986). *Actual minds, possible worlds.* Harvard University Press.

Bruner, J. S. (2002). *Making stories: Law, literature, life.* Harvard University Press.

Capobianco, B. M., & Yu, J. H. (2014). Using the construct of care to frame engineering as a caring profession toward promoting young girls' participation. *Journal of Women and Minorities in Science and Engineering, 20*(1), 21–33. https://doi.org/10.1615/JWomenMinorScienEng.2014006834

Ceci, S. J., & Williams, W. M. (2010). Sex differences in math-intensive fields. *Current Directions in Psychological Science, 19*(5), 275–279.

Ceci, S. J., & Williams, W. M. (2011). Understanding current causes of women's underrepresentation in science. *Proceedings of the National Academy of Sciences, 108*(8), 3157–3162.

Clark, M. C. (2010). Narrative learning: Its contours and its possibilities. *New Directions for Adult and Continuing Education, 126,* 3–11.

Cunningham, M., & Lachapelle, C. P. (2016). Designing engineering experiences to engage all students. *Educational Designer, 3*(9), 1–26.

Decety, J., & Jackson, P. L. (2004). The functional architecture of human empathy. *Behavioral and Cognitive Neuroscience Reviews, 3*(2), 71–100.

Diekman, A. B., Brown, E. R., Johnston, A. M., & Clark, E. K. (2010). Seeking congruity between goals and roles: A new look at why women opt out of science, technology, engineering, and mathematics careers. *Psychological Science, 21*(8), 1051–1057.

Dorie, B. L., & Cardella, M. (2013, June 23–26). Engineering childhood: Knowledge transmission through parenting [Conference paper 6966]. 2013 American Society for Engineering Education (ASEE) Annual Conference and Exposition, Atlanta, GA. https://peer.asee.org/engineering-childhood-knowledge-transmission-through-parenting

Eccles, J. S. (2005). Subjective task value and the Eccles et al. model of achievement-related choices. In A. J. Elliott & C. S. Dweck (Eds.), *Handbook of competence and motivation* (pp. 105–121). Guilford Press.

Eccles, J. S., & Wang, M. T. (2016). What motivates females and males to pursue careers in mathematics and science? *International Journal of Behavioral Development, 40*(2), 100–106.

Egan, K. (2012). *Primary understanding: Education in early childhood.* Routledge. https://doi.org/10.4324/9780203813577

Engineering Accreditation Commission. (2019). Criteria for accrediting engineering programs, Accreditation Board for Engineering and Technology (ABET) 2018–2019. www.abet.org/accreditation/accreditation-criteria/criteria-for-accrediting-engineering-programs-2018-2019/

Hess, J. L., & Fila, N. D. (2016). The manifestation of empathy within design: Findings from a service-learning course. *Codesign*, *12*(1–2), 93–111. https://doi.org/10.1080/15710882.2015.1135243

Hynes, M., & Swenson, J. (2013). The humanistic side of engineering: Considering social science and humanities dimensions of engineering in education and research. *Journal of Pre-College Engineering Education Research (J-PEER)*, *3*(2), article 4. https://doi.org/10.7771/2157-9288.1070

Katehi, L., Pearson, G., & Feder, M. (2009). *Engineering in K-12 education: Understanding the status and improving the prospects*. National Research Council, Committee on K-12 Engineering Education. National Academies Press. https://doi.org/10.17226/12635

Lachapelle, C. P., & Cunningham, C. M. (2014). Engineering in elementary schools. In S. Purzer, J. Strobel, & M. Cardella (Eds.), *Engineering in pre-college settings: Synthesizing research, policy, and practices* (pp. 61–88). Purdue University Press.

Letourneau, S. M., & Bennett, D. (2020, July–September). Using narratives to evoke empathy and support girls' engagement in engineering. *Connected Science Learning*, *3*(3). www.nsta.org/connected-science-learning/connected-science-learning-july-september-2020/using-narratives-evoke

Marra, R. M., Steege, L., Tsai, C. L., & Tang, N. E. (2016). Beyond "group work": An integrated approach to support collaboration in engineering education. *International Journal of STEM Education*, *3*(1), 17.

National Science Foundation. (2019). *Women, minorities, and persons with disabilities in science and engineering: 2019* (NSF Special Report 19-304). www.nsf.gov/statistics/wmpd

Peppler, K., Keune, A., Dahn, M., Bennett, D., & Letourneau, S. (2020). Designing for empathy in engineering exhibits. In C. Girvan, R. Byrne, B. Tangney, & V. Dagiené, (Eds.), *Exploring, testing and extending our understanding of constructionism: Constructionism 2020*, J80–81. Association for Computing Machinery.

Preston, S. D., & De Waal, F. B. (2002). Empathy: Its ultimate and proximate bases. *Behavioral and Brain Sciences*, *25*(1), 1–20.

Rusk, N., Resnick, M., Berg, R., & Pezalla-Granlund, M. (2008). New pathways into robotics: Strategies for broadening participation. *Journal of Science Education and Technology*, *17*(1), 59–69.

Su, R., & Rounds, J. (2015). All STEM fields are not created equal: People and things interests explain gender disparities across STEM fields. *Frontiers in Psychology*, *6*, 189.

Terenzini, P. T., Cabrera, A. F., Colbeck, C. L., Parente, J. M., & Bjorklund, S. A. (2001). Collaborative learning vs. lecture/discussion: Students' reported learning gains. *Journal of Engineering Education*, *90*(1), 123–130.

Walther, J., Miller, S. E., & Sochacka, N. W. (2017). A model of empathy in engineering as a core skill, practice orientation, and professional way of being. *Journal of Engineering Education*, *106*(1), 123–148. https://doi.org/10.1002/jee.20159

Walther, J., Miller, S. E., & Kellam, N. (2012, June 10–13). Exploring the role of empathy in engineering communication through a transdisciplinary dialogue

[Paper presentation]. 119th ASEE Annual Conference and Exposition, San Antonio, TX. https://peer.asee.org/exploring-the-role-of-empathy-in-engineering-communication-through-a-transdisciplinary-dialogue

Wigfield, A., & Eccles, J. S. (2000). Expectancy-value theory of achievement motivation. *Contemporary Educational Psychology, 25*(1), 68–81.

chapter 4

Co-Designing Learning Dashboards for Informal Educators

Elham Beheshti, Leilah Lyons,
Aditi Mallavarapu, Wren Thompson,
Betty Wallingford, and Stephen Uzzo

At the New York Hall of Science (NYSCI), researchers, technology developers, and museum practitioners participated in collaborative design, testing, and prototyping to jointly develop a digital "dashboard" that could be used to help facilitate our Connected Worlds exhibit. This enormous digital exhibit immerses visitors in a simulated ecosystem consisting of four biomes that they can manage through their body movements – for example, by gesturing in front of the exhibit's screens to plant seeds and prune away dead plants. The goal was to create a tablet-based dashboard that would provide updated live data from the exhibit. With this data, young on-the-floor museum staff, known as Explainers, could help visitors observe and make sense of key concepts and phenomena that emerged when visitors interacted with the exhibit, such as when a shift in water consumption upstream affected downstream flora.

This chapter details how a team of NYSCI researchers and technology developers worked with museum practitioners (Explainers, program facilitators, trainers, and managers) to co-develop a dashboard that could digitally monitor the state of the Connected Worlds exhibit, record trends, and flag interesting events in the simulated ecosystem. Our intent was to work with practitioners to figure out what information to present and how to present it so that Explainers could better facilitate the exhibit for visitors. This project had another important goal – to establish and study a design process centered on relationships among key participants and the agency that grows from these relationships.

In this chapter, we lay out the rationale and research foundations for our approach. We describe our design strategy, which emphasizes relationships among museum practitioners, researchers, developers, and others affected by the project. Finally, we discuss the strategies used to cultivate agency among the key participants in our research-practice partnership.

The Importance of Relationships and Agency in the Design Process

Creating new technologies for any workplace involves balancing pragmatic needs with innovative possibilities. On the one hand, existing work practices and relationships impose constraints on how a new technology should be designed. On the other hand, the technology makes wholly new work practices possible. Far too many workplaces are mired in timeworn practices that might be made more efficient, enjoyable, or beneficial with imaginative applications of technology. Because staff members often lack technological expertise, they may not be able to envision what might be possible with new technologies. At the same time, far too many developers think they know best and set out to build an innovative new technology that sadly goes unused because the developers failed to understand, or had no interest in understanding, how the technology would be integrated into the day-to-day practices of staff. In designing a digital dashboard for Explainers, we sought to develop a mutual understanding that would strike a good balance between the pragmatic and the possible. Toward this end, we used design methods that equalized the agency of key stakeholders, developers, and staff while maintaining the support of researchers.

Agency can be defined in many different ways (Matusov et al., 2016), from definitions that situate agency within the individual to those that emphasize the role of structures and systems in granting agency to individuals. In this project, we look at agency through a different lens that focuses on relationships. We emphasize the concept of "relational agency," which describes "how individuals . . . are able to engage with the dispositions of others in order to approach, assess and respond to the possibilities for action available to them in a situation" (Edwards & MacKenzie, 2005, p. 291). Relational agency is based on the relationships of individuals with their colleagues, with other groups of participants (such as staff members, researchers, and developers), and with levels of hierarchy in the system. In other words, our definition of agency combines elements of both the individual and structural definitions by emphasizing an individual's agency within their workspace relationships. Our definition further acknowledges that relationships are not static but are always evolving.

Applying this perspective of agency to the design process means that relationships need to be recognized and accommodated explicitly in the design methods. In our case, we created a version of participatory design, which we call "situated-relational design." It extends the concept of contextual design (Ahn et al., 2019), which addresses the situations in which work is done, by encouraging relational agency. We sought to accomplish this through processes such as mutual education, in which both practitioners (Explainers in this case) and researcher-developers help each other explore the available possibilities for action. Consistent with this definition of situated-relational, we identify different dimensions of designing a sociotechnical system in which people are the core element and their relationships are the main emphasis. These include existing and desired relationships not only with the technology but also with the hierarchy of people in both the workspace (such as program managers, Explainers, and visitors) and the design process (such as researchers, developers, and other participants).

Design Context and Challenges

Our project began as an outgrowth of a 1,000 square meter, six-story-tall, immersive digital exhibit called Connected Worlds (https://nysci.org/home/exhibits/connected-worlds/), which at the time of this study had been running for about five years. The first of its kind, Connected Worlds mixes virtual and physical reality to immerse visitors in a fanciful simulated ecosystem (see Figure 4.1). The simulation is projected onto six 4-meter-high screens, one 12-meter-high screen, and a projection floorspace of 214 square meters, all seamlessly melded into a continuous interactive environment. This ecosystem consists of four biomes (desert, grasslands, jungle, and wetlands) and three sources of water (waterfall, reservoir, and mountain valley). The simulated world is responsive to visitors' actions as they explore what it means to create a sustainable ecosystem.

The Connected Worlds exhibit allows visitors to manage the ecosystem by choosing how to distribute water flowing across the floor to different simulated biomes projected along the walls of the exhibit and selecting which types of and how many plants to plant in each biome. Because the water in the exhibit is finite, visitors have to decide how to share this resource by moving infrared-reflective stuffed "logs" on the exhibit floor to reroute the flow of water projected onto the floor. Visitors get very engaged in the water routing and planting, to the point that they often fail to notice the deeper phenomena presented by the exhibit (Mallavarapu et al., 2019). The Explainers who work at the exhibit reported difficulties with introducing visitors to ecosystem phenomena like the water cycle, ecological succession, and animal responses to habitats. They also face challenges in making visitors aware of phenomena related to complex systems, such as tipping points (when many small actions accrue until a sudden dramatic effect, like a massive die-off of plants, occurs) and tele-coupling (when an action in one location at one time produces an effect in another location at another time).

Figure 4.1 The Connected Worlds exhibit.

One reason why visitors may not notice these phenomena is because the exhibit is a complex system simulation – not all phenomena are present and observable at all times, and some may not be observable by all visitors in the exhibit. This is a far cry from a traditional hands-on exhibit like a swinging pendulum, which highlights a single phenomenon and has a single focal point, making it far easier for an Explainer to discuss with visitors. This challenge sparked our interest as researchers. We needed a way for Explainers, who are high school and college students, to better highlight and explain concepts and phenomena of interest, like tipping points, that emerged naturally as visitors casually interacted with the Connected Worlds exhibit.

We envisioned a dashboard that would allow Explainers to act as a "human-in-the-loop" – a person who is integrated into a dynamic system so they can guide and shape the system. The Explainer would be privy to data generated by the simulation and could use that data to help visitors make sense of the simulation and make choices about what to do next, thereby changing the simulation state.

Our prior work with co-designing technology for use by Explainers (Jimenez Pazmino et al., 2015) taught us that we needed to do more than just secure the participation of the end users in our process. We needed to situate our design and our design process within the larger sociocultural context of the museum (Lyons, 2018). This meant securing the participation of all tiers of the museum's interpretive staff, from front-line Explainers to top-level managers. It meant coming to understand the training experienced by the Explainers and the pedagogical philosophies that underpin that training. It meant reconciling the idealistic intentions behind interpretive practices with the actual ways those practices unfold in the messy real world and recognizing

elham beheshti et al.

how situational factors pragmatically affect the work. It meant developing techniques to solicit and listen to the needs and interests of our design partners. Moreover, we would need to develop techniques for helping our practitioner design partners, the Explainers, understand our research objectives for the project and the possibilities that data analytics and visualizations would offer for their practice. To engage them fully in the design process, we had to help familiarize our practitioner design partners with technical subjects that they may know nothing about.

Situated-Relational Design Process

We used four types of engagement – observations, focus groups, interviews, and participatory design sessions – to understand the design space and cultivate a productive co-design process. Each of these engagements involved mutual education: some engagements helped the researcher-developers understand practitioner needs and ideas, and others helped practitioners understand research goals and development possibilities.

The observations and focus groups were intended to help the researcher-developers understand Explainers' current practices and training practices and explore open-ended opportunities for technological support. The interviews helped the researcher-developers form a better picture of the relational agency experienced by practitioners in their workplace hierarchies. We used this information to conceptualize the tasks that practitioners currently perform, and would like to perform, during facilitation and to plan the participatory design sessions. These design sessions primarily helped the practitioners understand the possibilities and constraints of the technology being considered, while also supporting the collaborative production of designs (Beheshti et al., 2020). Below, we further describe each of the four engagement types.

The research team of four included two senior researchers from the museum with training in learning sciences and human–computer interaction (including participatory design methods), a computer science graduate student with experience designing data analytics and visualizations, and a junior researcher from the museum who had experience with leading usability studies and extensive experience working with the facilitators and the Connected Worlds exhibit. Over the course of the design process, we observed or spoke with 12 Explainers and two museum supervisors/trainers. We also worked intensively with four Explainers to produce and vet designs. We are currently working on creating prototypes of the dashboard that incorporate our findings from the design sessions. We will test our prototypes with Explainers in the exhibit after the museum reopens to visitors in 2021.

Observations of Practice and Professional Development

First, we conducted rounds of observations to study the Explainers at work during seven separate sessions over the course of three weeks. We used a partial interval recording sampling methodology (Frank et al., 2001), which kept track of whether certain behaviors occurred at any point during a specific time interval. Two researchers from our team took field notes of the behaviors and interactions of seven Explainers while they facilitated the Connected Worlds exhibit. We also observed two training sessions in which one of the managers of Explainer training, Uriah (a pseudonym), worked with 12 Explainers to review facilitation tasks and strategies. Two of our researchers took field notes and consolidated their observations after the training sessions ended.

Focus Group Study

To round out our understanding of the role and perspectives of Explainers, we conducted a focus group study with seven Explainers who had different levels of experience with facilitating the Connected Worlds exhibit. During the focus groups, the Explainers discussed in their own words what problems and challenges they faced when facilitating the exhibit, how they viewed the tasks involved in facilitation, and how these tasks might be supported by a tablet-based dashboard. The focus group structure was borrowed from the sociotechnical systems participatory design methodology (Jimenez-Pazmino et al., 2015), which helps participants imagine what their situation could be like socially, physically, and cognitively when using a new technological tool.

Interviews With Managers

Throughout the design process, we regularly met with the managers of Explainer training and exhibit experiences to gain a deeper understanding of the work of the Explainers. After we started our participatory design sessions (described next), we came to realize that we needed to gather more information about Explainers' practices, schedules, and training process, and the history of facilitation tasks around Connected Worlds. To get this information, we conducted a semi-structured interview of about two hours with one of the managers of Explainer training. We video recorded and transcribed the interview for further analysis.

Participatory Design Sessions

After completing the observations, focus group sessions, and preliminary interviews, we conducted a series of four participatory design sessions during which the practitioners and researcher-developers jointly generated design ideas for the "human-in-the-loop," data-driven dashboard. A dashboard needs to take into account three basic design elements:

- *Tasks.* The actions that users undertake while engaged in their practice
- *Scenario.* The situational contexts that need to be attended to
- *Data.* The data types, data transformations, and data visualizations that can provide information about the ongoing phenomena and processes

A reciprocal relationship exists among the tasks, the scenario, and the data, each informing the others. A developer familiar with analytics might be tempted to start with the data and consider the affordances and constraints of the data as a way of

elham beheshti et al.

shaping the scenarios and tasks, but we deliberately reversed this order. We began the process with the design element most familiar to our practitioners: the tasks they perform during their existing practice. We spread our design process across four different sessions to allow both practitioners and researcher-developers to arrive at mutual understanding as a means of promoting joint relational agency.

Table 4.1 summarizes the structure and activities in each participatory design session, and Figure 4.2 showcases the materials and activities used in these sessions. We conducted the study with four Explainers, each with more than three

Table 4.1 Structure of participatory design (PD) sessions

Session	Goals	Activities & Materials
PD 1 Task elicitation & brainstorming	Provide a situated focus on design space to specify tasks: *Goal:* Use a scenario to help distill the more generalized task ideas generated in the focus group (e.g., "help visitors understand the water cycle") into specific task enactments (e.g., "highlight for visitors when rain occurs")	*Materials* A printed scenario narrative that situates the Explainers within a story of use. Sticky notes (one color per participant) White board *Activities* Participants individually brainstorm tasks they might perform in the scenario, recording one task per sticky note Participants work together to sort tasks into related piles Participants label tasks that need technical support or training support
PD 2 Scenario generation & design sketching	Introduce research-driven constraints to reconceive scenarios and design interfaces: *Goal:* Explore how tasks can support collaborative learning goals *Goal:* Introduce the technology platform to be used (tablets) to adapt task ideas to work with a tablet while preserving their core task function	*Materials* Sticky notes (one color per participant) Papers with printed tablet outlines *Activities* Participants sort tasks from prior session by how well they support learning goals and add new tasks as needed Participants generate scenarios that support learning goals via tasks Participants generate paper designs for tablet interfaces to support tasks enacted within scenarios

(continued)

Table 4.1 Cont.

Session	Goals	Activities & Materials
PD 3 Data analytics generation & prototyping	Introduce data-driven constraints to reconceive interface designs and the tasks they support: *Goal:* Center proposed designs around the nature and availability of data drawn from the exhibit *Goal:* Engage practitioners in thinking about the ways data can be transformed to yield information *Goal:* Engage practitioners and researcher-designers jointly in creating paper prototypes that employ visualizations of transformed data	*Materials* Cards printed with available data variables Cards printed with possible data transformations Sheets illustrating common types of data visualization Papers with printed tablet outlines Worksheets to guide design process *Activities* Participants draw random data variable card to kick off design Participants select other data variable cards needed to support design idea and data transformation cards Participants select data visualization types that would be applicable Participants sketch out design Participant teams present joint design Participants rotate group membership, then repeat
PD 4 Early testing & prototyping	Introduce situational constraints to reconceive interface designs and the tasks they support *Goal:* Allow participants to explore how prototype design would fare in practice *Goal:* Shift design focus to address a research project goal (using live data for formative purposes rather than historical data for reflective purposes)	*Materials* Functional prototype of design derived from ideas generated in prior design session Papers with printed tablet outlines *Activities* Role play activity with functional prototype (one practitioner acts as facilitator, other participants act as visitors) Paper design activity inspired by insights gleaned during role play

years of experience. Each session was video recorded to capture the interaction of the participants with each other and with the materials. The conversations were recorded, and these audio recordings were later transcribed and thematically coded for further analysis. Each of the four sessions was 2.5 hours long, and they were conducted over a period of six months at NYSCI.

elham beheshti et al.

Figure 4.2 Examples of participatory design session materials and activities.

In the next section, we present an example of how one design idea, the Pokédex, evolved and how the design process supported relational agency and mutual understanding.

Illustrating Relational Agency: The Pokédex Case

We embarked on this design research with the goal of making an explicit break from prior approaches that treat the end users of educational technology as consultants and leave the nuts-and-bolts creation of technical designs to the "experts" (e.g., Martinez-Maldonado et al., 2016). In the absence of a relationship that invites and sustains ideas from both end users and researcher-developers, these experts run the risk of accidentally extinguishing creative ideas because of a seeming mismatch with research priorities. The joint design of the Pokédex feature illustrates a different approach based on relational agency that involves both end users (the Explainers) and researcher-developers.

Pokédex refers to the Pokémon Index, an encyclopedia of the characters from the widely popular Pokémon video game. This idea comes from popular culture and is not directly connected to any of our project goals, the learning goals for the exhibit, or the technological platform and its affordances or constraints. It is exactly the sort of idea that would be shrugged off by researcher-developers in a design process that did not support two-way relational agency. It is common for designers who are not experts to draw on designs they have previously used that may have only surface similarity to the design challenge at hand, simply because they are familiar references. But in a design process where practitioners are just as free as researcher-developers to suggest possibilities for action, the idea for a Pokémon Index evolved naturally and organically throughout the design sessions.

Brainstorming and Eliciting Tasks

The Pokédex idea originated in the focus group sessions when Explainers were discussing how they would approach visitors who seemed confused and needed help interacting with the exhibit. Lev (a pseudonym), one of the Explainers, made this point:

> I think what will make [starting a conversation] easier is if we had something like a mini version of the Living Library [an interactive guide mounted on a large screen in the entrance space of the exhibit] on an iPad, because there is so much in [the exhibit]. Yeah, just small overviews of [the exhibit] on the iPad.

We noted the idea, but since it was not necessarily aligned with our research goal of using data-driven visualizations as formative feedback in real time, we did not consider it an important design idea worthy of further exploration.

Soon after we started our participatory design sessions, however, we realized that the Explainers repeatedly returned to the idea. The need for "showing all the animals and plants" in an encyclopedic format got its Pokédex name during participatory design session 1 (PD1). This session had a strong task focus. When we asked the Explainers to collaboratively create a series of tasks for facilitating the exhibit and label them as "desired" or "existing" tasks, they wrote a card for "seeing all the different animals and plants, like a Pokédex," and labeled it as a desired task (Figure 4.3). When we asked them to group their task cards into buckets, they concluded that the Pokédex task did not belong to any one bucket, so we speculated that it would gradually disappear in subsequent design sessions. However, we were soon proved wrong.

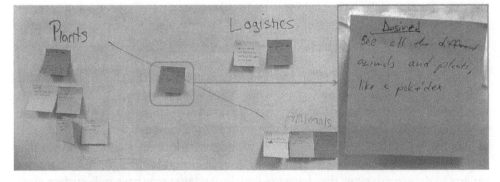

Figure 4.3 Task labels created by Explainers in the first participatory design session that express the desire to see and show all plants and animals in one biome.

elham beheshti et al.

Generating Scenarios and Sketching Designs

In the second design session (PD2), the purpose was to explore how the tasks generated in PD1 would change when we reemphasized research-driven constraints on the scenario of use, such as the overall learning objectives of the exhibit, the goals for collaborative engagement, and the technological platform to be used. We first asked the Explainers to revisit the tasks they had identified in the first session and create a bucket of tasks that would help to promote collaborative interactions among visitors. We were surprised to see the Pokédex idea reappear, as one Explainer suggested that a Pokédex interface could prompt collaboration. She explained that the desire to "collect" the different types of animals could give visitors a goal that would encourage them to work together to make the animals appear. Later, another Explainer pointed out that some visitors become confused when they cannot plant everything in all biomes. Having access to a Pokédex interface could help visitors understand which possible actions are applicable in an open-ended space.

At the end of PD2, we asked Explainers to choose a task from those they had previously generated. We asked them to imagine designs for an interface that would support a scenario in which that task could be enacted and to sketch their design idea. One of the design ideas was to show "silhouettes of animals and plants" that would be hidden at first and would therefore not be available for interaction. When a particular plant or animal appears in a biome, its silhouette would be "unlocked," enabling the users to tap on the picture and learn more about the plant or animal. This unlocking feature was proposed as a means of motivating visitors to work together to summon animals. It is drawn directly from the video game source of the Pokédex idea, and it would never have occurred to the researchers on the design team.

By this point of the design process, we had come to understand that the Explainers were not just proposing to duplicate a familiar and popular digital media element but were creatively drawing inspiration from the Pokédex idea to address both authentic needs that they repeatedly experienced in their work and the design goals we were proposing. We learned that in their regular practice, the Explainers have a paper "cheat sheet" with information about different animals and plants in each biome. Although the sheet was supposed to be a reference for their own use, they also employed it as a visual aid to stimulate conversations with visitors. (This practice diminished, however, as the sheets became wrinkled or lost, and as Explainers mastered the sheet's content.) The Pokédex idea was an extension of this authentic practice that could preserve and enhance the sheet's function of mediating conversation.

Our interview with Uriah, one of the managers of Explainer training, supported our observations from the PD sessions about the Pokédex idea. When asked his opinion of the core facilitative tasks, Uriah suggested that "talking to the visitors about what they're seeing and what they're noticing" is the priority. He added:

> We always tell the Explainers to be open to visitors finding their own connections . . . and maybe it's not the clouds going up into the air and moving over to the water, but it's the way that one bird goes over to a space and interacts with this plant, and then another bird appears.

As this comment shows, Uriah views helping visitors notice the animals and plants as a core facilitation task that can foster more meaningful connections with the exhibit. It also shows how managers can encourage relational agency between Explainers and visitors – the Explainers are encouraged to follow the lead of visitors' interests.

Generating Data Analytics and Prototyping

The third session, PD3, prompted a re-evaluation of the task and dashboard design. To help the Explainers understand the opportunities and constraints of the live data that would become available through the project, we created several physical props for the session (see Figure 4.4). We printed cards with the names and short descriptions of the variables in the simulation, such as the total amount of water in the system, the water present in each biome, the number and types of plants and animals, and the number and location of clouds. We printed cards with the names and descriptions of the transformations that could be performed on the data, such as calculating averages, ratios, and frequencies. In addition, we supplied sheets printed with examples of common data visualizations, including line graphs, bar charts, and pictorial representations, taken from The Data Viz Project.[1] We also supplied a worksheet to guide participants in connecting these design materials. After randomly selecting a variable card, participants would brainstorm how to use that data to engage visitors by constructing a scenario and how to transform and create visualizations of that data to make facilitation easier. These ideas could lead to the design of a prototype dashboard.

PD3 used a paired co-design model to encourage sociotechnical designs that more fully integrated the possibilities emerging from the marriage of technology design and practice. Each practitioner was paired with one researcher-developer at a separate table, and the pair brainstormed activities and created paper prototypes and

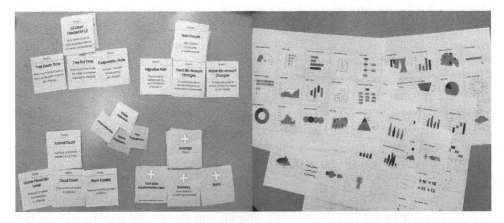

Figure 4.4 Materials for participatory design session 3, including cards representing available types of data and data transformations and printed examples of common types of data visualizations.

elham beheshti et al.

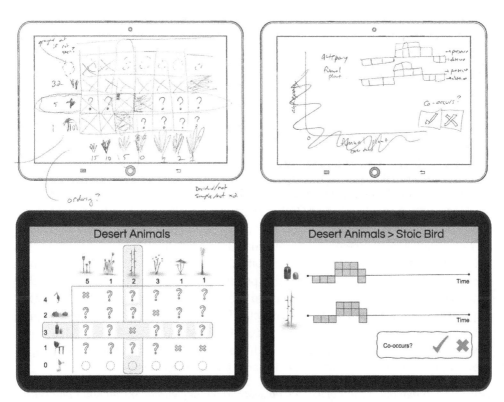

Figure 4.5 Initial rough sketches of a Pokédex-inspired interface (top) and a clearer illustration (bottom). Users could explore which pairs of plants and animals occur simultaneously in the same biome to determine whether a particular plant plays a role in forming an animal's habitat.

sketches of tablet screens. After a design cycle, one of the partners rotated to produce new practitioner-research pairs. This setup allowed both the researcher-developers and practitioners to share their expertise, much as the "jigsaw" classroom technique empowers and encourages students with different areas of knowledge to share expertise (Aronson et al., 1978).

In one of the cycles, a practitioner-researcher pair randomly picked the "count of animals" as the primary variable for their design. They brainstormed the design requirements and generated a prototype of a screen that shows in real time the animals that appear after, or as a result of, a plant appearing in each biome (see Figure 4.5). Interestingly, this design serves many of the same purposes as the Explainers' original Pokédex idea. At its core, it still documents different animals and their habitat needs and could be used to support Explainer-visitor conversations, but the way in which information is accessed has been completely revised. Rather than serving as an index for looking up static information about an animal, the interface is structured around an interactive task that allows the Explainer and visitors to explore together the possible connections between animals and plants, a task that could be used to encourage collaboration among visitors.

The Pokédex idea evolved because of the relational agency experienced by both the practitioners and the researcher-developers. Each found possibilities for action in the responses of their design partners in PD3, which led to an accommodation of their respective goals and ideas – the design equivalent of the "yes, and" strategy used by improv performers.

Early Testing and Prototyping

In PD4, the emphasis shifted again, this time by reintroducing a situational focus. Participants were asked to consider how it feels to use the dashboard when embedded in live interpretation. Does the dashboard make it easier to perform interpretive tasks that are not feasible in regular practice? Using an origami wireframing tool (https://origami. design/), we created a prototype of a non-Pokédex design produced in PD3 that focused on monitoring and engagement activities that Uriah had highlighted as a challenge for visitors. We tested the prototype in the exhibit using a role play activity in which one practitioner acted as the Explainer and the other practitioners acted as child visitors. After this testing, we moved to a lab space and asked the Explainers to design paper prototypes inspired by insights gained during the role play. Even though the prototype we designed was focused on real-time water usage and counts of live plants, the Pokédex idea found its way back into the discussions during both the enactment and the design activities, as shown by this conversation among participants (all pseudonyms):

Lev: *I am imagining . . . sort of like the Pokédex [laughing] with all the different plants and animals that could be there . . . so like the screen that we had there before [pointing at the tablet with wireframes] where Karl was using [it] to show the plants in the jungle . . . and [as I was playing the role of the kid] I wanted to see possible plants and animals that we can have.*

Karl: *So you want to have pictures of all plants and pictures of all animals.*

Lev: *Yeah.*

Karl: *Would you like to have something that correlates these two, like this animal comes out with this plant?*

Lev: *I think it could be one of them, like we could add different filters that could show it. I don't think it's a necessity.*

Karl: *I agree.*

Lev: *What do you think?* [asking Dhanashri]

Dhanashri: *I want to see the migration of the animals.*

Lev: *Oh, yeah.*

This vignette illustrates yet another evolution of the Pokédex idea. Rather than structuring it as a data-driven collaborative activity like the design that emerged during PD3, or a mystery challenge as suggested in PD2, or a simple index as conceptualized in PD1, here the Pokédex is accessed situationally. For example, when displaying information about the jungle region of the exhibit, Explainers wanted the option of showing which plants and animals could live in that region and the connections between them.

This case illustrates how our understanding of the Pokédex idea co-evolved as we applied different lenses to the interpretation design space. We began with a strong focus on tasks, shifted to a research-framed focus on a scenario, brought in a data focus, and then vetted the ideas through situated role play, refining participants' design ideas all along. By alternating the emphasis on the necessary design elements, participants could develop a design to a certain point before being asked to evaluate it through another lens. This process is akin to tempering a steel blade, in which alternating heat and cold strengthens the final product. These shifts in focus allowed all parties to examine and re-evaluate design ideas in a way that preserved relational agency. Under a different participatory design structure, the researcher-developers may have explicitly or implicitly asserted authority over the process to dismiss the Pokédex idea early on. In our participatory design structure, the process itself provided the needed vetting – the researcher-developers did not have to directly critique the idea because participants "kicked the tires" of the idea in the themed sessions. The practitioners kept finding creative ways to keep the Pokédex idea alive, adapting it to suit each session's focus.

Through this mutual education process, the practitioners came to understand the development constraints and priorities of the researcher-developers, and the researcher-developers came to understand the importance of centering the interface around the topics that visitors gravitated toward. This allowed the participants to develop versions of the idea that aligned with the needs of both practitioners and researchers-developers. We are currently working on a design that merges aspects of all suggested Pokédex designs. We plan to test this design, along with other prototypes that resulted from our design sessions, with our Explainers in 2021.

Cultivating a Relational Research-Practice Partnership

A key challenge in developing tools for open-ended learning activities is the need to attend to a variety of situational factors, such as the characteristics of the learning environment, the types of activities, and the hierarchy of relationships between learners and educators. All of these factors impact how a tool is used in the real world. Meeting this challenge requires more than just a design process like the one detailed above. It also requires a well-built partnership between the researcher-developers and the practitioners throughout the design process.

We believe that the need for a cultivated research-practice partnership goes beyond the objectives of a single research project. The longer-term goal is to create and maintain a sustainable relationship between the practitioners (our Explainers, program facilitators, trainers, and managers) and the researchers, content developers, and technology designers in an institution. (In cases like ours, the designers and developers are also the researchers.)

This section looks at the kinds of relationships that need to be considered in the research and design process to create sustainable tools and enable long-term partnerships.

Designing for an Evolving System of Relationships, Not a Static User

As commonly practiced, participatory design focuses on engaging the direct users of the solution in the design process, meaning that the design/research activities are centered around how the users interact with and use the solution. In our case, the users of the dashboard tool are Explainers, who are considered the main participants in the design process. However, our explorations and observations of Explainers' tasks, work schedules, and relationships to their trainers and supervisors revealed a tightly integrated work system at the museum (Lyons, 2018). This, in turn, led us to closely attend to the hierarchy of practitioners (Explainers, trainers, and supervisors) and how that hierarchy evolves (Fischer et al., 2008). Consequently, our research model engages not just the front-line practitioners (the Explainers) but also includes their trainers and supervisors throughout the design process. This approach allows us to integrate perspectives, insights, and buy-ins from the practitioners, who are not the immediate users of our design but play a key role in training, managing, and changing the Explainers' practices with the support tool.

For example, we learned from our interview with Uriah that the managers first wanted to assign Explainers with more than one year of experience to facilitate Connected Worlds, which is one of the most challenging exhibits to facilitate in the museum and requires a range of conceptual engagements and physical interactions with visitors. It is also one of two exhibits in which managers wanted to deploy new interpretive practices that were a focus of Explainers' training; these practices are part of a larger shift away from the didactic "explaining and giving information" approach toward more inquiry-based co-investigations with visitors. However, the pool of more experienced Explainers was too small, and, therefore, the managers had to eventually assign trainees – newly hired Explainers with very little experience – to these exhibits. The trainees are expected to work in pairs with a Senior Explainer who has more than three years of experience, but based on our observational studies, scheduling needs do not always allow for this. We also learned that since many of the trainees are high school students who typically come to the museum on weekends, they mainly facilitate sessions with family visitors. (The majority of weekend visitors are families while weekday visitors consist largely of school groups.) This means that novice Explainers can be tasked with facilitating Connected Worlds without having a full understanding of the details of the exhibit or an optimal level of experience with inquiry practices.

These insights made us more sensitive to the needs of the Explainers with less experience, who require greater support in learning the exhibit content, and the needs of all Explainers for continued professional development in facilitative practices. It also helped us understand that we need to think of our design as a tool that can seamlessly provide support to both new and more experienced Explainers by allowing them to layer in more complexity when they are ready to do so. Like the printed cheat sheets, the dashboard can be used by novice Explainers to learn about the exhibit and become more comfortable with initiating visitor conversations on their own. Unlike the cheat sheets, the dashboard covers more than just content; it embeds practice into its design, allowing even expert Explainers who know the exhibit content but are newer

elham beheshti et al.

to inquiry practices to gain experience with engaging visitors in joint investigations of subjects like animal migrations. Had we not developed an understanding of the evolving relationships between Explainers and the exhibit, we would have missed the opportunity to create a dashboard that will be useful and sustainable.

Establishing Mutual Trust and Understanding by Building History

Our process model encourages multilevel engagement of practitioners. Aligned with this model, we attended to the hierarchy of practitioners from day one of our design process by meeting with one of the managers of Explainer training (Uriah) and the manager of exhibit experience (Tarnvir, a pseudonym). We talked with them about Explainers' work as they engaged with Connected Worlds and other exhibits and programs at the museum. By keeping in touch with the managers from early in the process, we not only gathered information about how the exhibit is managed and facilitated, but, more importantly, we built trust and a history between us as researcher-developers and them as practitioners. This growing trust made it possible for us to collect data in a more structured way during our participatory design sessions by interviewing Uriah and engaging him in our design decisions.

We believe that continually keeping in contact with the managers created a structure for eliciting feedback in an authentic way. For example, before the scheduled interview with Uriah, two researchers joined him as he conducted morning training sessions. Once we started the interview, this joint experience served as an icebreaker that established common ground for a deep conversation. It demonstrated that we are not just interested in designing a dashboard tool for the purpose of research, but that we want to support the practitioners' work and care about their authentic needs. Moreover, by regularly contacting the managers, we showed them our commitment to the project and our respect for their process. In other words, our consistency sent a message that we do not want our partnership to disrupt or ignore their work but rather to collaboratively help improve the process.

Conclusion

This project recognized that agency is created and cultivated through an ecosystem of individuals and their relationships, rather than through single individuals. In our museum, this ecosystem consists of interconnected, evolving social interactions among visitors, Explainers, trainers, supervisors/managers, and researcher-designers. When research and design is conducted in this setting with the aim of developing a product that satisfies both the goals of researchers and the needs of practitioners who will use it, an inherent tension exists. The most important goals are not necessarily the same for researchers and practitioners, and compromise is required. We argue, however, that these competing goals actually produce a frisson that can generate design ideas that might never have emerged from one camp or the other. The challenge is to channel that frisson in a productive direction.

We make the case for using participatory design processes that bring to the surface and respect relational agency. Acknowledging relational agency makes space for different types of participants, who have different understandings and agendas, to build off of each other's ideas. During a participatory design process that promotes relational agency, a mutual understanding develops that yields innovative and useful designs that go beyond current best practices or current vogues in research.

Acknowledgments

This material is based on work supported by the National Science Foundation under Grant Nos. 1123832 and 1822864. Any opinions, findings, and conclusions or recommendations expressed in this material are those of the authors and do not necessarily reflect the views of the National Science Foundation.

Note

1   https://datavizproject.com.

References

Ahn, J., Campos, F., Hays, M., & DiGiacomo, D. (2019). Designing in context: Reaching beyond usability in learning analytics dashboard design. *Journal of Learning Analytics, 6*(2), 70–85.

Aronson, E., Blaney, N., Stephan, C., Sikes, J., & Snapp, M. (1978). *The jigsaw classroom*. Sage.

Beheshti, E., Lyons, L., Mallavarapu, A., Wallingford, B., & Uzzo, S. (2020, April). Design considerations for data-driven dashboards: Supporting facilitation tasks for open-ended learning. In R. Bernhaupt, F. Mueller, D. Verweij, & J. Andres (Chairs), *Extended abstracts of the 2020 CHI Conference on Human Factors in Computing Systems* (pp. 1-9). Association for Computing Machinery.

Edwards, A., & Mackenzie, L. (2005). Steps towards participation: The social support of learning trajectories. *International Journal of Lifelong Education, 24*(4), 287–302. https://doi.org/10.1080/02601370500169178

Fischer, G., Piccinno, A., & Ye, Y. (2008). The ecology of participants in co-evolving socio-technical environments. In P. Forbrig & F. Paternò (Eds.), *Engineering interactive systems* (pp. 279–286). Springer.

Frank, M., Steuart, T., & Christopher, H. (2001). Functional behavioral assessment: Principles, procedures, and future direction. *School Psychology Review, 30*(2), 156.

Jimenez Pazmino, P., Slattery, B., Lyons, L., & Hunt, B. (2015). Designing for youth interpreter professional development: A sociotechnologically-framed participatory design approach. In M. Umaschi Bers & G. Revelle (Chairs), *Proceedings of IDC 2015: The 14th International Conference on Interaction Design and Children* (pp. 1–10). https://doi.org/10.1145/2771839.2771840

Lyons, L. (2018). Supporting informal STEM learning with technological exhibits: An ecosystemic approach. In F. Fischer, C. E. Hmelo-Silver, S. R. Goldman, & P. Reimann

elham beheshti et al.

(Eds.), *International handbook of the learning sciences* (pp. 234–245). Routledge. https://doi.org/10.4324/9781315617572

Mallavarapu, A., Lyons, L., Uzzo, S., Thompson, W., Levy-Cohen, R., & Slattery, B. (2019). Connect-to-connected worlds: Piloting a mobile, data-driven reflection tool for an open-ended simulation at a museum. In *Proceedings of the 2019 CHI Conference on Human Factors in Computing Systems (CHI '19)* (pp. 1–14). Association for Computing Machinery. http://dx.doi.org/10.1145/3290605.3300237

Martinez-Maldonaldo, R., Pardo, A., Mirriahi, N., Yacef, K., Kay, J., & Clayphan, A. (2016). LATUX: An iterative workflow for designing, validating, and deploying learning analytics visualizations. *Journal of Learning Analytics*, 2(3), 9–39. https://doi.org/10.18608/jla.2015.23.3

Matusov, E., von Duyke, K., & Kayumova, S. (2016). Mapping concepts of agency in educational contexts, *Integrative Psychological and Behavioral Science*, 50(3), 420–446.

Part II
Relinquishing Power and Authority in Informal Settings

Part II

Museum–Community Engagement to Support STEM Learning

Andrés Henríquez and Marcia Bueno

When you pay attention to the community, the community responds with engagement, questions, and commitment. When you shift the culture of a museum to work with the community and provide real service and help, you create relevance with that community, which allows you to listen, learn, and continually improve. This is the foundation for the NYSCI Neighbors Initiative at the New York Hall of Science (NYSCI).

As one of the largest cultural institutions in Queens, NYSCI has a unique role to play in its immediately adjacent neighborhood of Corona, Queens. Corona is a culturally diverse community that is often a first stop for newly arrived immigrants from Central and South America and from East and Southeast Asia (Lobo & Salvo, 2013). Nearly two-thirds of the community is foreign-born, and more than 90% speak a language other than English. More than one-half of families in the Corona/Elmhurst district have incomes below 200% of the federal poverty level, a greater share than in the rest of Queens (Citizens' Committee for Children, 2019). Many hold multiple jobs to make ends meet.

Corona also has many community assets: a commercial hub filled with locally owned businesses, ethnic restaurants, and community-based organizations providing a range of services, as well as many cultural institutions and a perceptible cultural pride. Among the community's greatest assets are the highly aspirational values of Corona's immigrant families and their engagement in their children's education. NYSCI staff who work with families have observed parents routinely attending Parent Teacher Association (PTA) meetings at district elementary schools.

NYSCI provides informal learning experiences to a half million students, teachers, and families each year. For several years, we have been committed to building deep, long-term relationships with our local community and strengthening engagement of local families in museum visits and program activities, particularly those focused on science, technology, engineering, and mathematics (STEM) learning. We realized, however, that we needed a more concerted, museum-wide effort to become deeply integrated with our immediate neighborhood and better attuned to our community's needs and its unique cultural, business, and social assets.

Over the past four years (2016–2020), the centerpiece of our efforts to improve museum–community engagement, particularly for those who are immigrants and first-generation, has been an aspirational initiative called NYSCI Neighbors, developed in partnership with the local school district and nonprofit organizations. This initiative has developed an ambitious vision for using the museum's resources to (a) build an active community where students, teachers, families, and community members feel welcomed and empowered to engage in creative STEM learning, and (b) understand and address the social and cultural barriers to community engagement in STEM opportunities. This initiative also has a research component that seeks to learn from emergent practices in museum studies and to define and test specific strategies for engaging immigrant-origin children and their families with STEM educational opportunities and career pathways.

In this chapter, we describe NYSCI's vision, strategies, and programs for two-way community engagement with a focus on STEM learning, as well as the impacts to date. We also share challenges and lessons learned about building stronger relationships between museums and the diverse families in their immediate communities.

Why Museum–Community Engagement Is Crucial

At its core, community engagement recognizes that children learn in a complex ecosystem that extends well beyond formal educational settings (National Research Council, 2009). Informal learning settings like museums and science centers play a unique role within this ecosystem.

What Museums Can Offer

Museum and science centers like NYSCI have unique advantages that enable them to serve as alternative sites of opportunity for STEM learning and provide experiences

andrés henríquez and marcia bueno

that deepen and extend what schools and families have to offer. They are unrestricted by the testing and accountability mandates that govern public schools. Therefore, they have the freedom and flexibility to better capitalize on the interests and assets of diverse and immigrant families and codevelop creative STEM learning experiences that are more accessible and relevant (Connolly & Bollwerk, 2017; Karp et. al, 1992; Norton & Dowdell, 2016; Simon, 2010). They already have diverse and engaging offerings for STEM learning, deep practitioner knowledge, and a potential to respond to the needs of the students and community.

Museums and science centers can also play a broader role in sustaining communities, especially in ethnically diverse communities with many immigrant families and substantial educational, economic, and social needs. For example, through strategic partnerships with other community institutions and organizations, museums can help to address community needs in a coordinated way. The benefits of these efforts are mutual, in that they help to ensure the vitality of both museums and communities.

With the COVID-19 pandemic, Corona and other nearby neighborhoods emerged as the epicenter of New York's raging outbreak. This has created dire needs in NYSCI's local community that will make community engagement even more critical.

Particular Need for Engagement in Diverse Communities

In communities with a diverse population and many immigrant families, there are often cultural barriers to engaging with museums. Cultural values and perceptions of museums and STEM vary among different groups. Children of immigrants who are likely to choose STEM fields tend to be of Asian descent, and recent research shows that Latinx and Central American immigrants are less likely to expose children to "cultural capital" related to STEM learning, such as going to a science museum, and are less likely to encourage their children to take higher-level math courses in high school or to pursue a STEM college degree (Ma & Lutz, 2018). For many of the immigrant families we serve, their first visit to NYSCI is their first time in any museum.

Immigrant families are also affected by educational disparities that create barriers to engagement but also make engagement with informal learning opportunities especially important. Children of immigrants attend both public and private schools with widely varying economic resources, linguistic services, and student–teacher characteristics (Suárez-Orozco, Abo-Zena, & Marks, 2015). Many immigrant students attend schools that struggle with the quality of their pedagogy and are located in communities with serious risks to students' well-being and achievement potential (Suárez-Orozco, Hernández, & Casanova, 2015). Low socioeconomic status combined with high ethnic composition in a school can have a negative impact on students' academic performance (Suárez-Orozco et al., 2018). Research by Suárez-Orozco et al. (2008) found that many immigrant children "attended schools that not only obstructed learning and engagement but also were, in many ways, toxic to healthy learning and development" (p. 89). In particular, schools attended by immigrant students are often highly segregated, not only by color but also by income level and language, isolating their high-need students from the greater school resources and academic expectations that are typical of well-funded schools.

In NYSCI's local community, most children attend Community School District 24, one of the city's most crowded districts, with 56 schools serving over 60,000 English language learners and children from low-income families who receive federally funded Title I support (New York City Department of Education, 2017). The district also routinely performs below the state average in math and English language arts tests. Local families are understandably concerned about disparities in resources and educational opportunities for their children. Compounding these disparities is the lack of spaces where children and families can learn together after school. As an assistant principal told us, "There is simply nowhere for our kids to go [after school]."

Fulfilling this potential requires active, two-way engagement and well-designed programs. NYSCI Neighbors was created to address this need. Through this initiative, we aspire to serve as an anchor institution that offers a safe and welcoming space for families. We are marshalling the museum's assets to provide greatly needed resources and opportunities at no cost for thousands of local children and their families, particularly immigrant families. We have also incorporated systematic inquiry into our collaborative improvement efforts in order to identify strategies and lessons that can inform the engagement work of other cultural institutions.

NYSCI's Approach to Community Engagement

Several foundational research studies on immigrant communities and STEM learning have shaped our approach to designing community engagement efforts.

"Ecologies of Care" Framework

The work of Carola Suárez-Orozco, an advisor to this initiative and a leading researcher on immigrants' identity formation, family separation, civic engagement, relationships with schools, and other experiences, provided a theoretical framework that has informed NYSCI's community work and guided the development of our concrete programs and strategies for engaging families. Her "ecologies of care" framework consists of structured practices designed to support immigrant children in adapting to a new education system and society. Originally developed to address issues within schools, this assets-based framework recognizes that immigrant students bring their own agency, resiliency, and important cognitive strengths to the table, while also acknowledging the challenges these students face in their new environment. It involves the active effort to offer educational opportunities that are responsive to students' needs and have the potential to create an ethical cycle of engagement among students who feel cared for (Noddings, 2015; Suárez-Orozco, 2018). NYSCI has adapted this framework in its community STEM work by focusing on students' participation in after-school enrichment programs including academic support and mentoring, facilitation of parental involvement, and activities to increase families' explicit knowledge about college pathways (Suárez-Orozco, 2018).

andrés henríquez and marcia bueno

Design Make Play Approach to STEM Learning

The NYSCI Neighbors initiative, like our other initiatives and programs, is premised on our Design Make Play pedagogy, which is anchored in decades of research about how people build deep understandings of STEM concepts and practices (Falk & Dierking, 2016; Honey & Kanter, 2013; National Academies of Sciences, Engineering, and Medicine, 2018; National Research Council, 2009). This approach is designed to give learners more agency to construct knowledge, build on prior experiences, and investigate personally and socially meaningful problems. It is characterized by open-ended exploration, imaginative learning, personal relevance, deep engagement and delight, discovery, and problem-solving. (See this book's Introduction for more details about the Design Make Play principles.) Consistent with Design Make Play, learners have opportunities to interact with skilled facilitators who ask good questions and provide feedback and guidance, which has been shown to enrich learning (Robertson et al., 2015; Windschitl, 2002).

Strategies for Community Engagement

The NYSCI Neighbors initiative combines the community framework and learning approach described above to build bridges to Corona families and a wide range of community stakeholders who might otherwise not actively choose to experience NYSCI's exhibits, programs, and products. Below, we discuss the specific strategies we used, with examples of the activities we offered, to increase engagement in creative STEM learning opportunities and related goals.

Making Institutional Commitments for Deeper Engagement

In order to maintain and increase trust and engage deeply with families and their children, we made institutional commitments to support engagement programs, broaden our knowledge about Latinx immigrant families, and refine our support system and learning approach. For example:

- To support a greater understanding of the local community and culture, we hired staff who were raised in and live in the local community.
- To build awareness of NYSCI in the community, we hired a community coordinator to maintain relationships with key stakeholders and develop programs that would build awareness of NYSCI's offerings. Parent coordinators, who work in the schools, became essential partners in this work.

A Focus on the Family

Although we understand that a community encompasses more than the families who live in it, much of our work focuses on families as the basic unit for engagement. Many of our programs, including those described in Chapters 6 and 7 of this book, are premised on an intergenerational approach in which children and their parents or other caregivers engage in STEM learning together (Shonkoff & Fisher, 2013). This approach recognizes that when parents are involved in their children's learning, children are more engaged, learn more, and over the long term are more likely to attend school and graduate. Our approach also seeks to leverage the high value that many immigrant parents place on education and the central role of families in shaping their broader communities.

This emphasis on family further acknowledges that many parents in Corona are conversant only in their native language, largely Spanish, and rely on their children to do a lot of interpretive work for them. However, these parents can be empowered by experiences that provide opportunities for them to engage in hands-on STEM educational experiences with their children and to learn the same STEM concepts and practices as their children are learning.

An Emphasis on Building Trust

Many parents may be intimidated going into a museum, starting with the large, imposing physical structure. Parents need to feel trust that NYSCI is a safe environment where the entire family is welcome and where they can interact with people who speak their native language and will make them feel at home. Our interviews with families, described later in this chapter, confirm that parents want an accessible, comfortable environment where they don't feel out of their depth, including activities that provide a clear starting point for visitors who don't have much knowledge about STEM.

We had learned that families felt comfortable in institutions like their local schools and church but were less familiar and comfortable with the idea of a science museum. Although NYSCI has been in the community for over 50 years, many local families had never been to the museum. It was a delicate process to ensure our families felt welcomed into NYSCI and that they experienced the museum as a safe space. Many of the strategies and activities discussed later in this chapter are designed to build families' trust and agency so they feel comfortable in the museum. One step we took was inviting the community into the museum on a regular basis for various activities geared to the community at large, such as English as a Second Language (ESL) classes, yoga classes, and community celebrations. In addition, NYSCI staff took advantage of opportunities to go out and be visible in the community as part of PTA meetings, pop-up events, and school events.

Two-Way Communication and Relationships

Regardless of the knowledge and experience of museum staff, engagement does not exist by merely inviting community members into the museum and designing programs for them. True engagement requires listening to and engaging in dialogue with community members about their needs, desires, aspirations, and challenges. It includes using their responses as the basis for codesigning programs.

We learned this from our prior work with area schools and community-based organizations to engage more directly with and support Corona's children and their

andrés henríquez and marcia bueno

families. In 2010, NYSCI began a program of work with a handful of schools to design an education program to tackle issues of summer learning loss. These schools purchased low-cost memberships for their families to participate in NYSCI's summer program, which included bilingual orientation sessions to familiarize parents and children with the museum's offerings and simple worksheets that encouraged children to explore the museum's exhibits while completing required learning assignments.

While this work seemed promising and was effective in drawing local families into the museum, we wanted to build on its success and create a deeper and richer set of opportunities to engage families from our neighboring communities. We learned from this previous work that providing access is not enough, especially to engage families who may feel like the museum is "not for us" or do not see a clear connection between the learning that happens at NYSCI and the skills and competencies that can help their children succeed in the future.

To develop a richer and more targeted initiative, NYSCI started by conducting in-depth interviews with community residents and leaders to better understand what they might want from an informal learning institution and what they saw as the community's education needs. Using focus groups, informal interviews, observations, and walkabouts with key stakeholders in the community, we explored these questions: How did parents navigate the K-12 school system? What career aspirations did they have for their children? What were residents' perceptions of NYSCI, and what key barriers kept them at arm's length from visiting the museum? How can NYSCI be more helpful to them? Participants in this process included parents, educators and school district administrators, members of the local clergy, and community-based organization staff. Specifically, we worked with 21 parents, 10 school principals, 5 school district administrators, a representative from the Community Education Council, 3 coordinators of after-school programs in schools, 4 managers of after-school programs in community-based organizations, a member of the clergy, and the community outreach personnel at the Queens Museum and Queens Zoo. Their responses provided a basis for the NYSCI Neighbors initiative.

Forging Community Partnerships

To support programming that was outside of NYSCI's areas of competence, we engaged in partnerships that could offer the kind of skill-building that families were looking for, such as family literacy, ESL, and health and wellness classes. We actively sought out partnerships with groups and organizations that focused on Latinx families and immigrant issues.

When NYSCI pivoted in response to COVID-19, we used the partnerships developed through NYSCI Neighbors to become an anchor for the community. We delivered virtual programs (such as Making at Home and Literacy through STEM). NYSCI was the lead in launching the Elmhurst/Corona Recovery Collaborative, a group of 22 nonprofit organizations working together to address and reach across a broad array of issues facing our community.

Consistent with these strategies, NYSCI's community engagement work is multilayered and bidirectional. Museum-centered programs are coordinated with school- and community-centered programs. We designed NYSCI Neighbors to connect our other outreach programs and activities that build relationships with families, schools, and local organizations and partners in Corona.

Together, these strategies form the foundation of a design approach that has led to more dynamic ways for NYSCI to engage with families and their children.

Community Engagement Activities
Responding to Community-Identified Educational Priorities

Parents participating in our interviews and focus groups said they wanted NYSCI to be a place that welcomed the entire family for intergenerational learning. Immigrants we interviewed as part of this project also indicated that, other than the local library, there were no places where families and children could learn together. A recent report from a child advocacy group noted the needs of parents and caregivers for homework help and college preparation and the general lack of quality and availability of educational resources (Citizens' Committee for Children, 2019).

To be responsive to families' priorities, NYSCI developed free after-school programs at the museum for students and families who attend schools that are part of NYSCI's network of community schools. These programs included homework help sessions and intergenerational STEM-focused activities. Key examples of these activities are briefly described below. (As of May 2021, the museum was closed due to COVID-19 and scheduled to reopen on July 2, 2021.)

- *Homework help.* Free homework help sessions are held Monday through Thursday from 3:00 to 3:45 p.m. for ten weeks every year. NYSCI staff and local high school students offer basic academic support to students as they work on their homework. Comments from caregivers suggest that this program is appreciated and is having an impact. For example, one caregiver who did not attend school as a child and has lived in Corona as a Mexican immigrant for many years was looking for bilingual resources in the community that could help her eight-year-old granddaughter, Abigail (a pseudonym), who had moved to the United States from Mexico in 2018. Upon hearing about NYSCI from a parent coordinator, the caregiver said, "I was always wondering what's in that museum." She noted that after learning more about the museum exhibits and programs, she immediately registered Abigail. After a year of participation in the homework help zone, she indicated that Abigail's English had improved so much that she was put in an English-only class.
- *Family STEM learning.* Multigenerational learning at the museum is a keystone to family engagement and parent empowerment. After the homework session, caregivers and their children are invited to explore the museum from 3:45 to 5:00 p.m. and participate together in a variety of hands-on, playful STEM activities (see Figure 5.1). These include interacting with the Connected Worlds immersive, simulated ecosystem exhibit described in more detail in Chapter 4; tackling engineering design challenges in NYSCI's Design Lab; using tools and

　　　　　　　　　　　　andrés henríquez and marcia bueno

Figure 5.1 A parent and her child engaged in a design activity.

everyday materials to make artifacts in NYSCI's Maker Space, as described in Chapter 7; and participating in live bilingual science demonstrations. Each session is attended by approximately 25 families. A total of 450 families and their children have participated in this program over the last two years. After their participation, families are asked to reflect on these experiences to help museum staff gain a deeper understanding of how the museum can add substantial value to their children's lives.

Parents participating in our interviews and focus groups further noted that they wanted to be informed about the educational value of what their children were doing at NYSCI – how the activities connected to STEM subject matter and skills like problem-solving and critical thinking. Our parent workshops were a response to this stated need.

- *Parent workshops.* NYSCI offered parents a set of free workshops and learning experiences designed to help them understand and reflect on how they can use hands-on, playful activities and discussion at NYSCI to explore STEM ideas with their children. These sessions were held on Friday afternoons at NYSCI for ten weeks every year. NYSCI staff encouraged parents to keep a personal journal to capture their reflection and help guide their thought processes during each session. Staff used journal prompts to guide parents' reflections: "What did this exhibit make you think of? Draw a picture of how it makes you think." The journals also helped parents make each activity and exhibit more relatable to their everyday lives through questions like these: "Do you see connections to this in your homes or in your everyday life?" A reflection time was held after each session where parents could connect the theme of the day to their children's lives and schoolwork and their own past experiences.

Empowering Parents

In addition to family programming at NYSCI focused on academic learning and STEM exploration, we worked with local partners to design activities to empower parents as leaders and advocates for their children in their broader community. During the design process, we focused on ideas for intergenerational activities to address three issues that the community deemed critical: (a) navigating a complex education system, (b) increasing parents' understanding of science and making informed decisions for their families, and (c) celebrating their cultures and networking with community partners.

- *School transitions.* Navigating the New York City K–12 school system is complex even for the most involved families. For immigrant families that are undocumented it is doubly challenging. Parents and students get support from caring teachers and a network of well-informed parents, but they also need expertise on navigating the system and choosing schools that are supportive of English language learners. To respond to these challenges, NYSCI developed critical school transition resources and activities attuned to families' language and culture. These activities were aimed at providing parents with access to information and people that can help them navigate the New York City school system and access opportunities available in STEM-focused high schools, colleges, and careers. We hosted high school and middle school fairs where families could learn more about school options and the complex high school application process in New York City (Figure 5.2). In our last two years, 80 students from Corona participated in the program. We also hosted six career awareness sessions in the evenings during the school year to introduce young people to possible careers in health and medicine, design and engineering, women in STEM, and conservation and energy. To date, 1,344 students have participated in these events, including 336 from our local community. We hosted 53 speakers. A total of 31 collaborating organizations, ranging from college programs to corporations and out-of-school time partners, contributed to the content of these events.

NYSCI partnered with several community organizations to design adult-focused courses and workshops at NYSCI to address critical needs, such as ESL classes and health and wellness, and to help parents better support their children's educational needs. As noted above, these activities were also intended to build trust and make families comfortable in the museum environment. Examples include the following:

- The Hispanic Federation offered workshops on pathways to academic excellence, early childhood literacy, and pathways to college.
- The New York City Department of Education's Office of Adult and Continuing Education offered two three-hour classes a week at NYSCI in STEM and ESL.

Figure 5.2 High school STEM fair for community youth.

- The Plaza del Sol Family Health Center at Elmhurst Hospital shared their literacy, nutrition, and wellness workshops with parents, and the Queens Zoo offered ten weeks of nature yoga classes to parents.

Language Accessibility and Multicultural Programming

A key strategy for building trust is to ensure that activities respect, represent, and celebrate the language and culture of the local community. Our programs for parents are all done in Spanish, and our science lessons include materials and activities that families are familiar with and have easy access to. Offering instruction in their native language engages families and allows them to be more open to science.

We have also learned the importance of being relevant to the community and recognizing the linguistic and cultural assets and "funds of knowledge" in local families and the community (Bang & Medin, 2009; Moll & González, 2004). Cultural recognition and celebrations are essential ways to show respect for families, have fun, and share an experience where the museum is a part of the community.

- *Multicultural events.* Since 2017, we have held several multicultural events to show appreciation for Corona families and our community partnerships and provide opportunities for social networking. NYSCI organizes a community festival each fall where residents showcase their expertise and talents. The festival has welcomed more than 1,000 visitors and highlighted our local cultures (Mexico, Ecuador, and the Caribbean) with dance and musical performances. In addition, every fall we invite all our local principals, parent coordinators, and the Community School District superintendent to a lunch to celebrate the beginning of the new school year. NYSCI participates in the Annual Community Appreciation Day, which is an opportunity for us to celebrate our community and give recognition awards to adults (parents and grandparents) from the community who have used NYSCI to advance the learning of children in the community.

Through all of the strategies described above, we have engaged an average of 3,000 local families per year in museum-based programs.

Impact of Community Engagement Activities

To understand the impact of this work, we began in 2018 to conduct interviews with family participants at the museum site. The interviews were semi-structured and were conducted in pairs, with two caregivers at a time. The interviews were mostly in Spanish and lasted approximately half an hour. The general topics explored included their overall satisfaction with and enjoyment of the programs they have participated in, how they perceive their children's progress in school, their comfort in coming to the museum, and the number of events they have attended. Because many of the community and family engagement programs have been underway for a relatively short time, we have not yet assessed the specific educational outcomes and science learning of participating students.

These parent interviews indicated both successes and challenges of the program. Overall, parents had a unified view of how the program benefited their children's education. They perceived the program as improving language ability for both their children and themselves, improving children's academic performance at school, and developing healthy study habits for their children, such as completing homework on time and maintaining a longer attention span while studying. They recognized the educational value of the experience. Parents also shared that the experience provided quality family time and helped them bond with their children. The following are some of the specific impacts reported by participants:

- Participants came to see NYSCI as a place for them, their families, and their community. Parents viewed the museum as a community resource, not just a science museum.
- In interviews and reflection sessions, participants reported high levels of satisfaction with the program. They indicated that they appreciated the STEM learning taking place, the home-school connections being forged, and greater confidence in supporting their children, as captured by these comments:

Participant 1: *[My child] was very timid and now he's not. So that's why I'm also letting him stay in school until 4 p.m. I saw the teacher yesterday and she said that he's changed. Now he participates!*
Participant 2: *I like all of it because it helped, for example, with the homework that they leave them sometimes or things that can be decorated at home.*

andrés henríquez and marcia bueno

- Parents said they have gained knowledge of and confidence in explaining exhibit contents and design activities to their children, as suggested by this comment:

  I have been able to understand what each section of the museum is about because we have visited before . . . it is what best helps me.

- Parents reported that the time spent with their children was valuable in and of itself, as indicated by comments like these:

  Parent 1: *To spend time with the kids more than anything . . . when we are here, it's specifically to be with them or to learn things together.*
  Parent 2: *It was good because I'm able to go out with the kids and not be just [indoors] or on the phone all the time. You learn a lot. I really liked it.*

- Additionally, the program is successfully activating new participants for NYSCI; in each new cohort, one to two enrollees tell us that they learned about the program from past participants. Several participants have taken the course for the second time, and about two families per cohort have indicated interest in continuing learning through a deeper program.
- Parents increased their capacity to navigate the New York City Department of Education and advocate for their children as a result of skills and knowledge gained in NYSCI's partners' offerings, such as increased English language abilities, workshops on decoding individualized education programs (IEPs), and high school and college preparedness workshops.
- A select group of parents are becoming leaders in the initiative, doing personal recruitment and recommending the program to friends. As the number of the initiative's alumni grows, we are moving closer to having a parent council and continuing to invite parents to share their experience of program participation at PTA meetings and similar venues.

In the future, we plan to assess education impacts such as changes in parents "funds of knowledge" in science, their comfort in advocating for their children in school and navigating the educational system, and milestones in children's STEM education.

Lessons Learned and Challenges

Our work to date has yielded a number of lessons that continue to inform the design and development of our STEM community programming:

- In engaging immigrant-origin parents, providing access to resources is only the first step. Institutions must go further, creating experiences that acknowledge community needs and, specifically, parents' needs.
- The intergenerational focus is key to making parent engagement possible. Attendance and participation are enhanced when parents can bring their children, and parents relish opportunities to learn alongside their children.

- At events focused on educational transitions (e.g., high school or job fairs), parents are supportive but let their children take the lead.
- Local families place a high priority on learning, but for many, their views of learning and education are traditional and academic. Parents expect to see workbooks, homework, drills, and other forms of study they themselves may have been exposed to in their education. Therefore, parents don't always recognize all NYSCI activities as learning.
- Engaging immigrant and first-generation parents with novel approaches to STEM exploration takes time. We have the most success when we first invite parents to pursue new learning experiences alongside their children, and then invite them to reflect on what they experienced and observed.

We have also identified some overarching and unforeseen challenges to working with this immigrant community. We are using these observations to drive changes to the design of the program.

- Building a relationship of trust and recognition is a challenge. A program that is for the community needs to be much more visible in the community.
- The neighborhoods we serve have a tremendous amount of transience. This means that we need to continually work at making NYSCI visible, communicating our offerings, and building new relationships.
- In-school learning is parents' first priority. As a result, getting families to identify and value the engagement that happens at NYSCI as learning is more challenging than anticipated. In the interviews, many parents only focused on the educational benefits of the homework help, rarely mentioning other parts of the program or NYSCI activities. This suggested that their view of learning and education is still centered around formal education, and that they had not adopted the open-ended, exploratory approach to learning that NYSCI advocates. Solving problems through design and making did not register with parents as an educational benefit of the program, even when their children were deeply engaged in this type of activity.
- Parents also raised several administrative issues in terms of program facilitation, schedules, and communications, and provided suggestions to address these issues.
- Finally, we face an overarching challenge in balancing our responsiveness to the community on issues such as immigration and health with our limited capacity to deliver programs outside our expertise of STEM engagement and learning in informal environments. While we have learned that it is necessary and valuable to understand families' many needs, our actual knowledge and capabilities lie in science experience, education, and learning through play.

From Local to National – The Coalition for a STEM Future

NYSCI has taken the lead in testing and broadening the community engagement model described above through a new national partnership of museums called

andrés henríquez and marcia bueno

the Coalition for a STEM Future. The Coalition is composed of five museums, each of which is working in its local setting to support first-generation immigrant families: NYSCI, Explora (Albuquerque, NM), The Exploratorium (San Francisco, CA), Houston Children's Museum, and The Tech Interactive (San Jose, CA). These museums are invested in learning how to leverage their resources in partnership with community stakeholders to use STEM as a pathway of opportunity for children of immigrants. The goal is to aggregate effective practice at the local level. The resulting frameworks and best practices will allow us to amplify our collective impact and help to inform policymaking and public and private investments.

NYSCI serves as a hub for the Coalition, whose work is experimental and in its early stages. The network of museums is using a Network Improvement Community (NIC) model of collaboration. We brought these like-minded institutions together to work toward defining and implementing common goals in rigorous and coordinated ways, while also being responsive to the unique circumstances and opportunities in their local sites (Bryk et al., 2011). NICs allows organizations to both improve practices and build capacity for continued improvement and collaboration with other organizations that have shared goals but are pursuing them in widely different ways (Bryk et al., 2015; Deming, 2000). This promising work has much potential to help build museums' capacity to work with communities.

Conclusion

Being relevant to the community in which we are located is critically important to families and their children, especially for immigrant communities. Using the approach and strategies described in this chapter, we have evolved a body of work that respects the cultural assets of the community and creates learning opportunities that respond to local needs. We are continually finding ways to be more responsive and attuned to the community's needs and engage more effectively with communities. Through our community engagement efforts and family programs, we are providing opportunities for families to learn about STEM concepts through a hands-on rather than a traditional academic approach, pursue their aspirations for their children's future, become familiar and comfortable with NYSCI, and grow as leaders and STEM advocates in their community.

Our analysis of community and family engagement initiatives thus far has provided information that we have used to inform our current program design, identify critical issues, and shift resources to where they are most needed. It has also helped us review our capacity to conduct research activities on this program.

In the future, we will begin to formalize program refinements that address the challenges of trust, turnover, and the tension between formal and informal learning. We will do so by designing more engagements that take place at school and community events, developing more personalized communication with schools, finding ways to integrate school programming into the museum, developing more drop-off programs, and continuing to broaden and increase communication about what happens at NYSCI.

Acknowledgments

We are grateful to those who have helped to support much of NYSCI's community work in Corona: Science Sandbox, an initiative of the Simons Foundation; Carnegie Corporation of New York; and Deutsche Bank Americas Foundation. Any opinions, findings, and conclusions or recommendations expressed in this material are those of the authors and do not necessarily reflect the funders' views. We would like to thank the committed teachers and parent coordinators in the schools who have worked with us over the years. Of course, we would be nowhere without the support of our families, who have encouraged us in so many ways over the years and have been champions of this work, even becoming informal ambassadors for our program.

References

Bang, M., & Medin, D. (2009). Cultural processes in science education: Supporting the navigation of multiple epistemologies. In L. D. Dierking & J. H. Falk (Eds). *Science learning in everyday life* (pp. 1009–1026). Wiley Press.

Bryk, A. S., Gomez, L. M., & Grunow, A. (2011). Getting ideas into action: Building networked improvement communities in education. In M. T. Hallinan (Ed.), *Frontiers in the sociology of education* (pp. 127–162). Springer.

Bryk, A. S., Gomez, L. M., & Grunow, A. LaMahieu, P. (2015). *Learning to improve: How America's schools can get better at getting better.* Harvard Education Press.

Citizens' Committee for Children of New York. (2019). Elmhurst/Corona Queens: Community drive solutions to improve family and well-being. https://buff.ly/2BzxzFw

Connolly, R. P., & Bollwerk, E.A. (2017). *Positioning your museum as a critical community asset: A practical guide.* Rowan and Littlefield.

Deming, E. W. (2000). *Out of crisis.* MIT Press.

Falk, J. H., & Dierking, L. D. (2016). *The museum experience revisited.* Routledge.

Honey, M., & Kanter, D. (2013). *Design, make, play: Growing the next generation of STEM innovators.* Routledge.

Karp, I., Kreamer, C. M., & Lavine, S. D. (Eds.) (1992). *Museums and communities: The politics of public culture.* Smithsonian Institution Press.

Lobo, P., & Salvo J. (2013). *The newest New Yorkers: Characteristics of the city's foreign-born population–2013 Edition.* New York City Department of Planning. www1.nyc.gov/assets/planning/download/pdf/planning-level/nyc-population/nny2013/ac.pdf

Ma, Y., & Lutz, A. (2018). Jumping on the STEM train: Differences in key milestones in the STEM pipeline between children of immigrants and natives in the United States. *Research in the Sociology of Education, 20,* 129–154.

Moll, L. C., & González, N. (2004). Engaging life: A funds of knowledge approach to multicultural education. In J. Banks & C. McGee Banks (Eds.), *Handbook of research on multicultural education* (2nd ed., pp. 699– 715). Jossey-Bass.

National Academies of Sciences, Engineering, and Medicine. (2018). *How people learn II: Learners, contexts, and cultures.* The National Academies Press. https://doi.org/10.17226/24783

andrés henríquez and marcia bueno

National Research Council. (2009). *Learning science in informal learning environments: People, places, and pursuits.* The National Academies Press.

New York City Department of Education. (2017). DOE data at a glance. www.schools. nyc.gov/about-us/reports/doe-data-at-a-glance

Noddings, N. (2015). *Ethic of care and education.* Teachers College Press.

Norton, M., & Dowdell, E. (2016). *Strengthening networks, sparking change: Museums and libraries as community catalysts.* Institute of Museum and Library Services. www.imls.gov/sites/default/files/publications/documents/community-catalyst-report-january-2017.pdf

Robertson, A. D., Atkins, L. J., Levin, D. M., & Richards, J. (2015). What is responsive teaching? In A. D. Robertson, R. Scherr, & D. Hammer (Eds.), *Responsive teaching in science and mathematics.* Routledge.

Shonkoff, J. P., & Fisher, P. A. (2013). Rethinking evidence-based practice and two-generation programs to create the future of early childhood policy. *Development and Psychopathology, 25*(4, Pt 2), 1635–1653. https://doi.org/10.1017/S0954579413000813

Simon, N. (2010). *The participatory museum.* Museum 2.0. www.participatorymuseum. org/read/

Suárez-Orozco, C. (2018). Ecologies of care: Addressing the needs of immigrant children and youth. *Journal of Global Ethics, 14*(1), 47–53.

Suárez-Orozco, C., Abo-Zena, M., & Marks, A. (Eds.). (2015). *Transitions: The development of children of immigrants.* New York University Press.

Suárez-Orozco, C., Hernández, M., & Casanova, S. (2015). "It's sort of my calling": The civic engagement and social responsibility of Latino immigrant-origin young adults. *Research in Human Development, 12,* 1–16. http://dx.doi.org/10.1080/15427609.2015.1010350

Suárez-Orozco, C., Motti-Stefanidi, F., Marks, A., & Katsiaficas, D. (2018). An integrative risk and resilience model for understanding the adaptation of immigrant-origin children and youth. *American Psychologist, 73*(6), 781–796.

Suárez-Orozco, C., Suárez-Orozco, M.M, & Todorova, I. (2008). *Learning a new land: Immigrant students in American society.* The Belknap Press of Harvard University Press.

Windschitl, M. (2002). Framing constructivism in practice as the negotiation of dilemmas: An analysis of the conceptual, pedagogical, cultural, and political challenges facing teachers. *Review of Educational Research, 72*(2), 131–175.

chapter 6

Big Data for Little Kids

Developing an Inclusive Program for Young Learners and Their Families

*C. James Liu, Kate Maschak,
Delia Meza, Susan M. Letourneau,
and Yessenia Argudo*

> So that's the takeaway as a parent. It makes me feel like, okay, let my kids grow to whoever they want because it might lead them to other explorations and other opportunities. They are exploring and they are growing . . . it's been learning for him but also for me.
>
> Big Data for Little Kids parent participant

Children are growing up in a world shaped by the rapid growth of "big data" and data analytics. To navigate this environment, children need to develop the capacity to understand and communicate core data concepts, and their parents and caregivers need to be able to support their children in this process. Informal learning environments can play a critical role in helping children and their families learn about and reason with data. However, little attention has been given to understanding how museum exhibits and programs can be made more inviting to family groups that are underserved in science, technology, engineering, and mathematics (STEM) fields.

The Big Data for Little Kids (BDLK) program at the New York Hall of Science (NYSCI) addresses these needs for children and their families from groups underserved in STEM fields. Designed by a team of NYSCI educators, program developers, and researchers, the goals of this multi-week program are to help children aged 5–8 and their caregivers learn together about data concepts through hands-on activities and shared experiences and, more importantly, to foster caregivers' agency in supporting their children's learning. In this program, children and their caregivers design their own exhibits by making sense of and using data. They measure the size of museum exhibits using everyday tools like tape measures. They collect data, make graphs and charts, and use the data in designing their exhibits.

In this chapter, we summarize the research base that informed our development of the BDLK program. Next, we give a brief overview of the context, content, and structure of the program. We then share our approaches and lessons learned about data modeling and engaging families from diverse cultural backgrounds, using examples and feedback from participating families. Finally, we discuss the challenges and implications for future work.

Research Background

The BDLK program draws from a body of research about the role of caregivers in informal STEM learning, particularly those from underserved groups, and about strategies for helping young children learn about and work with data in formal and informal learning environments.

The Role of Caregivers and the Needs of Underserved Groups

Parents and other caregivers play an important role in shaping children's experiences in informal learning environments. Within science centers like NYSCI, caregivers can support their children's learning in several ways. They can connect the content presented by exhibits and programs with children's personal interests and prior experiences (Ash, 2003; Crowley & Jacobs, 2002; McClain & Zimmerman, 2014). They can help children progress toward greater understanding by having openended conversations and encouraging hands-on exploration (Callanan et al., 2017; Leinhardt et al., 2002; Haden et al., 2014). Despite a substantial body of research showing how caregivers can support their children's engagement in informal learning spaces, few studies have elaborated on how exhibits and programs can be designed to invite family groups that are underserved in STEM-related fields (Pattison & Dierking, 2018; Siegel et al., 2007). As a result, families with nondominant cultural and linguistic backgrounds are often marginalized in informal STEM learning environments, and their access to informal learning experiences remains limited (Dawson, 2014, 2019).

This challenge of supporting caregivers in informal STEM experiences may be particularly salient for exhibits and programs that focus on complex or newly observed scientific concepts, such as those that involve making sense of and reasoning with data. Caregivers may have had little or no opportunities to explore data with their children (Ben-Zvi & Garfield, 2004). Moreover, data may be presented in schools, museums, and other educational settings in ways that lack cultural and social relevance for underserved families (Philip et al., 2016). These factors may hinder caregivers' agency and confidence in their ability to support their children in learning to use data.

Educating Young Children About Data

A substantial body of research has been conducted in formal education on young children's early mathematics and data education. In particular, studies in classroom settings have shown ways to help children engage in the core practices of working with data using the framework of data modeling. In this framework, children generate questions that need to be answered by data, define meaningful attributes of data based on the questions asked, and organize and analyze data in order to make predictions and decisions, all with support from educators over a series of lessons (English, 2012; Lehrer & Schauble, 2002, 2004). Researchers in early mathematics education have also advocated for playful and learner-centered approaches that tap children's curiosity and personal experiences to engage them in using data to answer open-ended, real-life questions (Clements & Sarama, 2014; Greenes et al., 2004). Together, these findings suggest that for young children, early learning experiences working with data involve exploring basic concepts in mathematics and building on children's innate capacities for observation and inquiry (National Research Council, 2013). Educators and caregivers can support this type of learning by providing young children with a context for gathering and interpreting data to lay the foundation for more complex statistical reasoning (Konold et al., 2015).

Data Modeling for Young Children in Informal Learning Environments

Extensive research has identified effective strategies for engaging children in using data to deepen their exploration of topics that are relevant to their lives and prior experiences (English, 2012; English & Watters, 2005; Lehrer & Schauble, 2002, 2007). As this body of research highlights, when children, even at a young age, are given well-designed instruction and scaffolding support that helps them gradually progress to greater understanding, they are capable of grasping the core concepts of data modeling and applying these concepts to make sense of the world around them and reach decisions. Unfortunately, children's opportunities to learn about data are not only limited in early grades, but also typically confined to the mathematics classroom and taught in procedural and formulaic ways (Makar & Rubin, 2009; Pfannkuch, 2018), especially in underfunded schools (Jackson & Cobb, 2010; Schoenfeld, 2002).

Informal learning may be a particularly important avenue for the early development of data modeling concepts and skills. Research in both cognitive development and education has shown that well before children enter school, they develop basic scientific and mathematical reasoning abilities through their everyday experiences (Greenes et al., 2004; Gopnik et al., 1999; Schulz & Bonawitz, 2007). Studies also indicate that informal learning environments can play a critical role in supporting early scientific and mathematical reasoning and cultivating children's interests in STEM (Falk & Dierking, 2018; National Research Council, 2009; Rogoff et al., 2016). Nevertheless, there has been little research on how young children learn about relatively complex STEM topics like data modeling in informal settings, especially for family groups that are already underserved by a variety of educational experiences.

Program Development, Structure, and Implementation

The BDLK program was developed through a sustained collaboration between informal educators, program developers, and researchers at NYSCI. The team worked together to brainstorm ideas for the program, prototype hands-on activities with museum visitors, and iteratively improve the curriculum.

In the first stage of program development, we identified key data modeling and mathematical concepts that could be adapted into hands-on activities in a family program. Drawing from prior studies (Lehrer & Schauble, 2002; English, 2012; Konold et al., 2015), we settled on four key concepts of data modeling to be covered in the program:

- Understanding the purpose and process of using data to inform decision-making
- Exploring different types of data and different data collection methods
- Practicing organizing and representing data using different data formats and perspectives
- Making sense of datasets through data analysis and interpretation to inform decisions and generate new questions

In addition, the team aimed to provide opportunities in the curriculum for young children to apply and practice early mathematics concepts such as patterns, comparison, addition and subtraction, and geometry.

The next step was to ground the program in NYSCI's Design Make Play pedagogical approach. (See this book's Introduction for more details about the Design Make Play principles.) Consistent with the principles of Design Make Play, the BDLK curriculum aimed to invite children and their families to apply a wide variety of skills, knowledge, and experience to specific data-related tasks and to take ownership of constructing their own data learning experiences. This approach also allowed the team to deepen our understanding of how family-focused programs could foster inclusion and agency in STEM learning and to identify strategies and practices that could be applied in developing future programs.

Before finalizing the BDLK curriculum, the project team prototyped ideas for program activities with museum visitors in order to determine the program's overarching theme, workshop structure, and hands-on activities. We sought to provide an authentic experience in which children and their families could use data to explore questions that are personally meaningful and worth solving and design solutions that are divergent and creative. Since the program involved a wide age range of children, we needed to ensure that our planning and implementation was flexible and that our activities and tasks varied in complexity. This iterative prototyping cycle helped guide the development of the overall structure of the workshop and curriculum.

In the final version of the workshop, called Museum Makers: Designing with Data, children and their families were invited to design and build a model of an exhibit that they felt was missing from NYSCI by gathering data about museum exhibits. The workshop structure consisted of a weekly two-hour session for seven consecutive weeks. In the first five weeks of the workshop, families learned about the steps involved in data modeling and used them to gather different types of data each week. These included data on their favorite exhibits, the size of exhibits, the time visitors spent playing with exhibits, and the different features of exhibits (see Table 6.1). These topics intentionally covered a range of types of data that children could directly measure using simple tools like stopwatches, tape measures, and written tallies and records that they kept track of in their own notebooks.

Each lesson in the workshop began with an introductory hands-on activity in which children practiced using measuring tools in the classroom – for example, they timed themselves completing a puzzle or measuring their height with measuring tapes. This was followed by a brief introduction to the driving questions that would guide data collection that day, such as "Which exhibit is the biggest?" Families then gathered data about exhibits on the museum floor. They measured the size of exhibits, timed each other as they interacted with exhibits, and counted features of exhibits like lights,

Table 6.1 Activities and guiding themes of workshops

Week		Theme Questions	Activities
1	Introduction	What should your exhibit be about?	*Exploring two exhibitions in the museum.* Ask children to find one favorite exhibit and one that taught them the most.
2 & 3	Measurement (size)	How big should your exhibit be?	*Measuring sizes.* Families use different measurement tools to measure the size of exhibits on museum floors.
4	Measurement (time)	How long will people stay at your exhibit?	*Timing activities.* Families use a stopwatch to time how long people stay or interact with exhibits.
5	Feature	How does your exhibit work?	*Identifying features.* Families identify different features included in exhibits (buttons, switches, lights, etc.).
6	Making	How should you design your exhibit?	*Designing and making exhibits.* Families plan their exhibits and use materials provided to create them together.
7	Presentation		Children present their exhibits on the museum floor.

sounds, and text, among other activities. Next, families came back to the classroom to compile and organize their data and use it to answer the driving questions. Finally, children planned and reflected on their own exhibit ideas, revising their plans from week to week as they learned more about the existing museum exhibits based on the data collected. In the sixth week of the workshop, children created models of their exhibit with their parents. During the last week, children presented their exhibits in a final project showcase (Figure 6.1) that was open to museum visitors and staff. (See Donnelly et al., 2018, and Letourneau et al., 2020, for more program details.)

We piloted the workshops in early 2017 and implemented two rounds of the program later in 2017 and 2018 with two different cohorts of families. The participants included 15 families (seven in Cohort 1 and eight in Cohort 2), with 24 children (12 in each cohort). The average age for Cohort 1 children was 6.42 years old, and 58% of this group were girls; the average age of Cohort 2 was 6.58 years old, with 50% girls.

Families were recruited from schools in NYSCI's immediate community and surrounding neighborhoods in Queens, New York, one of the most diverse counties in the United States (Lobo & Salvo, 2013). Almost two-thirds of residents in our immediate neighborhoods were born outside the United States, most commonly in Central and South America and in East and Southeast Asia. More than 23% of households in these neighborhoods live below the poverty line (New York City Department of Health, 2015). Almost all participating families were first- or second-generation immigrants. Fifty-three percent of the participating families spoke Spanish, and 33% spoke Mandarin at home. Most of the children were bilingual, and some caregivers spoke little or no English. We had bilingual staff on-site throughout the workshops.

Figure 6.1 Children showcasing their exhibit prototypes with museum visitors and staff.

Findings About Data Modeling

Throughout the iterations of the program, we tried to better understand how children and their families engage with core concepts of data modeling. We also looked at which parts of the program were important for supporting children and their parents, which aspects were challenging for them, and what helped them overcome these challenges. The data collected consisted of interviews with children and caregivers throughout the program, video and audio recordings of families' conversations, field notes, and artifacts that the families created in the program, such as prototypes for new exhibits and data displays. We used a mixture of qualitative and quantitative methods to analyze the data. We coded families' conversations about data modeling concepts according to themes we identified, and then we quantitatively analyzed the coded data (English, 2012). We did the same kind of coded data analysis of parent–child interaction styles from Cohort 1 families (Pianta & Harbers, 1996). We also conducted qualitative data analysis using inductive approaches (Thomas, 2006) to summarize the patterns observed in the workshops, the emerging themes from interviews, and artifacts created by families from both Cohorts 1 and 2.

Our analysis of family conversations demonstrated that the workshops generated a high level of conversation that was relevant to the program's theme of engaging families in data modeling. Among all conversations recorded during active workshop periods, 52% of the statements made were part of conversations about one or more data modeling concepts, including asking questions about data, discussing attributes of data, and collecting, organizing, or interpreting data. (Table 6.2 shows examples of family comments about these concepts.) Most of the conversations were about collecting data. However, the percentage of conversations about collecting data decreased over the course of the program, from 57% in week two to 44% in week five, while the percentage of conversations about interpreting data increased over time, from 9% in week two to 35% in week five.

Taken together, these findings suggest that during the workshops, caregivers and children talked extensively about the processes of data modeling. Over the course of the workshop series, the parent–child interaction style also evolved in two important and related directions. First, caregivers became less directive in their exchanges with their children and used more guiding strategies, such as asking open-ended questions (see Figures 6.2 and 6.3). Second, parent–child conversations increasingly focused on making meaning of data (the interpretation and analysis stages of the data cycle).

Our qualitative data analysis further indicates that with caregivers' support and scaffolding, children made sense of the data they gathered in personally meaningful ways and used that information to inform their decisions while designing their exhibits. Moreover, children and caregivers often connected their prior knowledge of mathematics operations like addition and their personal experiences with the data tasks at hand. Similar to prior studies, these results suggest that, even at a young age, children are capable of grasping the core idea and practices of data modeling.

Table 6.2 Coding scheme for data modeling concepts in parent–child conversation

Category	Definition	Example
Identifying questions	Talking about questions that could be answered by gathering data or asking what the question is that they were trying to answer	"How long are his legs?" "What exhibition has most of the features? Which one has more of everything?"
Defining attributes of data	Talking about what types of data to collect to answer a question or what counts as data	"Let me see . . . so we can measure the fan, up to down." "You used three [types of measurements]: measuring tape, cubes, and this."
Collecting data	Talking about measuring, counting, gathering information	"I set the timer. Are you ready to play? I'm going to time. Ready? And go!" "The head equals . . . one pipe cleaner [long]."
Organizing data	Talking about ordering or arranging pieces of information, visualizing data, making a graph or chart	"I have to make the chart to organize it." "Where do our higher times go [on the timeline]?"
Analyzing / interpreting data	Talking about summarizing data, or drawing conclusions based on data	"The longest people stayed was 20 minutes, so which [exhibit] do you think people liked better?" "This, the bubbles [exhibit], has the most features. Right?"
Other mathematics concepts	Talking about other math concepts not related to data science concepts above (e.g., multiplication, geometry)	"This is two times two."

Findings About Engaging Families From Diverse Cultural Backgrounds

In addition to helping children and their families better understand the core concepts of data modeling, the project aimed to enhance our understanding of how to invite families from diverse cultural backgrounds to engage with museum exhibits and, more importantly, how to welcome and support their needs in family programming. To accomplish this, we carefully examined our existing practices and decided whether they met the needs of these families. This became particularly crucial as we quickly realized at the beginning of our program that the diversity of participating families was much greater than we initially expected. By no means could they be conceptualized as a homogeneous group of "underserved families." Families varied not only in their nationalities, ethnicities, languages, socioeconomic status, and cultural backgrounds, but also in their involvement in, value of, and approach to

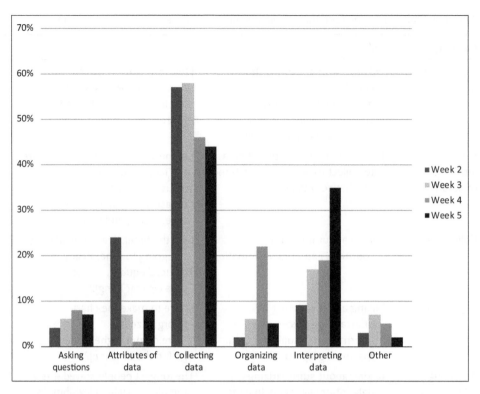

Figure 6.2 Percentage of caregiver–child talks devoted to particular data modeling topics.

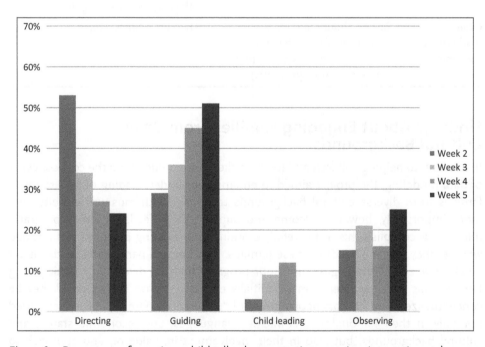

Figure 6.3 Percentage of caregiver–child talks demonstrating caregiver interaction styles.

c. james liu et al.

supporting their children's learning and development. Further, families differed in their prior museum experiences and expectations for the program.

In order to design a program that embraced this diversity, the team adopted an iterative research and design approach. We gathered feedback and information from the families through observations of program workshops, conversations and interviews with the families, and project team debriefs after each workshop. We synthesized and reflected on that information and used it to guide our program design.

We grouped our reflections into two categories, discussed below. The first category, overarching principles, represents our beliefs about how our core values might be enacted in an engaging family program. The second category, strategies, lays out specific guidelines and facilitation practices that helped us design and implement a program that embodies the overarching principles.

Overarching Principles

As we reflected on our practices, we recognized that certain core values were shared by participating families and our project staff. These core values are the most important elements of an engaging family program for all stakeholders. We conceptualized these core values as overarching principles that could be used to guide our process of creating and implementing the BDLK program. These principles emerged from three main sources: (a) our analysis of and lessons learned from the families' feedback; (b) the museum's overarching pedagogical approach of Design Make Play and our institutional initiative on creating a diverse, equitable, and inclusive learning environment; and (c) our project staff's years of experience with developing family programs and working with young children and their families. We believe these principles illustrate what an inclusive, engaging family program should look like from both our own and families' perspectives. We identified five overarching principles:

- Embrace, celebrate, and employ the range of diversities in our participating families
- Build a strong trust between the institution and the families
- Create a shared experience for the families and our museum staff
- Provide quality time for the families to be together
- Support both caregivers and children throughout this learning process

Below we discuss how we implemented these principles and what we learned.

Embrace, Celebrate, and Employ the Range of Diversities in Participating Families

We recognized that the first and most important principle was to go beyond simply respecting individual and family differences and take more active steps to embrace and mobilize the range of family diversity we found among our participants. We applied this principle as we developed the foundation of the program, cultivated the relationship between museum staff and caregivers and children, and established channels for communication. We see the values of diversity, equity, and inclusion as critical to our mission and essential to fostering a genuine sense of trust between participants and staff. These relationships create pathways toward elevating the learning capacity of every participant by understanding their diverse needs and providing modes of learning to fit those needs. Providing these opportunities has increased caregivers' engagement in and commitment to the program and has cultivated a sense of community.

Build a Strong Trust Between the Institution and the Families

In keeping with the first principle of embracing diversity, we took steps to create a two-way feeling of trust that would lead to shared ownership of the program. Despite caregivers' differing expectations for or experiences with museum programs, they trusted that our museum staff cared about their children's education and development and respected caregivers' own goals and expectations for the program. At the same time, the museum staff trusted and respected the ways in which families constructed their own learning experiences. Staff recognized that the experiences valued by families may or may not align with the intended learning outcomes of the program. Establishing this mutual trust set the foundation for the overall inclusiveness of the program by creating an expectation that the museum would listen to and respect families' perspectives.

Create a Shared Experience for the Families and Our Museum Staff

While this simple idea of creating a shared experience is the cornerstone of our family programs, it may be easily neglected or overlooked. In our program, the shared experience went beyond simply being in the same space or doing tasks together. Rather, we sought to create a sense of community to be shared by all of the children, all of the adults, and all of our staff who participated in the program. We did this by emphasizing that this was the first time that everyone involved was exploring the process of data modeling. We told the families that while this topic may seem intimidating, data modeling is new for us, too, so we would all be learning together. By establishing this common ground, caregivers felt comfortable expressing any questions and concerns and free to explore data in ways that made sense to them. They were encouraged to share their honest feedback about the program. Through this process, we were able to help and support each other and enjoy our time together without the pressure of working toward a single, predetermined "right" answer. The sense of belonging and collaboration motivated family members to be involved in and care about not only their own children's participation but also each other's experiences in the workshop.

Provide Quality Time for Families to Be Together

The feedback we received from caregivers and children throughout the workshops helped us to recognize the importance of providing "together time." When asked what they felt were the most important and memorable aspects of the program, many caregivers shared that the program provided dedicated and uninterrupted time for

c. james liu et al.

them to be with their children and to play, learn, and have fun together, as expressed by this quote from a parent:

> It is common knowledge we don't have a lot of time, so this hour with our children, it is giant for us, to be able to spend time with our children. About school, there is not a lot to talk about, except helping with schoolwork. But here, we come here, we do things together, and we talk about lots of things we have done.

This principle echoed the consistent findings from visitor research that spending quality family time is often the most important reason to visit museums and other informal learning environments (Falk & Dierking, 1992, 2018).

Support Both Caregivers and Children Throughout This Learning Process

In particular, we aimed to help caregivers to feel confident and comfortable learning about data with their children, as well as powerful and knowledgeable enough to support their children in working with data. In our program, many caregivers were unfamiliar with concepts such as data modeling and also new to the learner-centered approach embedded in the program curriculum. However, they quickly repositioned themselves as learners so they could support and share in their children's learning and reflect on their own learning experiences. By fostering caregivers' empowerment, ownership, and confidence and ensuring that they had a positive learning experience alongside their children, we saw that they emphatically enriched the quality and intent of the learning experience for their children and for us.

Together, we learned that an engaging, inclusive, and supportive family program depends on a shared and enjoyable learning community. This community is built on a trusted and positive social environment that encourages learners of all ages and backgrounds to be active and confident in their own learning.

Specific Strategies for Program Design and Facilitation

To achieve the principles described above, we identified a set of effective practices to help us develop and implement the program. We grouped these practices into four strategies:

- Providing a welcoming, accessible, and respectful learning experience
- Setting clear expectations for learning goals
- Inviting caregivers to be co-facilitators
- Supporting caregivers' needs

Providing a Welcoming, Accessible, and Respectful Learning Experience

Through our program, we learned that respecting and embracing every family's unique cultural needs and perspectives and making everyone feel comfortable were the key to creating an engaging and inclusive learning environment. This meant adjusting the program to meet families' needs rather than expecting families to change to fit the parameters of the program. We had to ensure that the content throughout the program was meaningful, approachable, and understood by all family members and that activities were inviting to both children and adults.

For example, although the program was specifically designed for children ages 5–8, it needed to cater to a much wider range of family members, from infants to grandparents. The program had to welcome families with a range of configurations and needs and accommodate other demands on caregivers' time. Instead of limiting participation to children in the target age and their caregivers, we looked for ways to engage the entire family group. We not only designed workshop activities that worked well for five-to-eight-year-olds, but also adjusted activities as needed to engage younger and older family members in their own ways. This strategy helped to invite all family members – parents, grandparents, younger siblings, and others – to be actively involved in the program together.

On the programming side, we needed to take a closer look at how the program welcomed and provided access to our local families. Even though NYSCI has worked hard to establish close relationships with local families, schools, and organizations, it was still difficult for the team to recruit families who were not regular museum visitors for a multi-week program. We had to rethink whom we were inviting and what we should learn from our communities. As mentioned above, we realized that we needed to allow families to decide who would come along – for example, whether they would bring young siblings. We also needed to allow different caregivers, whether a mother, father, or grandparents, to participate from one week to the next depending on their other obligations. Finally, we needed to adjust the workshop schedule to allow families to use the museum as a space to meet after school before going home. These details were crucial to ensuring that the program was inclusive and welcoming for our local families.

Setting Clear Expectations for Learning Goals

In order to invite all family members to join the program, we found it extremely helpful to set expectations for families about what the program was and what they would experience. This paved the way for staff to emphasize and clarify the learning goals of the program for families, especially the caregivers. In particular, we explicitly invited the families to focus on the process and not become overly concerned about the content in the activities. This was especially helpful for caregivers who were unfamiliar with our learner-centered approach and whose prior experience with STEM instruction had been more traditional and didactic.

Setting these expectations helped them to move beyond just seeking out "right" answers in the activities and to focus on constructing knowledge by exploring, collaborating, and making sense of what they were experiencing with their children. This also helped caregivers to engage younger children, who were just starting to develop complex mathematical concepts, in the activities and to find connections through children's own perspectives and experiences. One caregiver appreciated the opportunities for creativity inherent in this approach:

For my child, [the program] really allows [his] creativity to come out, and not just that, it also allows parents to be part of it . . . It's nice that we learn more about our children and we help them learn a little bit more.

In an activity known as Measuring James, for instance, children used tools such as cubes, pipe cleaners, and measuring tapes to determine the size of an outlined figure called James. Children could choose which parts of the figure they wanted to measure – for example, the width of the head or the length of the legs – and which of the aforementioned tools they would use to do the measuring. While older children were able to see how different units of measurement affected their data, the younger children, with help from their caregivers, were able to focus on identifying what to measure and collect data using whatever tools they chose. The open-ended quality of the activity allowed children and adults to find ways of interpreting the task that felt developmentally appropriate, interesting, and meaningful to all members of the family.

Inviting Caregivers to be Co-facilitators

Our team identified several strategies to invite caregivers to be actively involved in the learning process with their children throughout the program. For example, the team used various channels to better communicate with everyone in family groups. We aimed to eliminate language barriers for families – some children and caregivers were English language learners, and some families included bilingual children and caregivers who did not speak any English. This latter situation can marginalize caregivers by forcing them to rely on their children as interpreters. Direct solutions included translating all printed learning materials and instructions and having staff available who spoke different languages. We recognized, however, that reducing language barriers required more than translations. It was also important to build personal relationships with the families by getting to know them. This created an environment in which families felt safe and encouraged to ask questions and share their ideas and stories with us and other families. For example, we shared a light meal with families at the end of each day of the workshops. This setting naturally encouraged everyone to converse about the workshops and their lives and prior experiences.

We also learned that it was critical for staff to directly communicate with caregivers, not just with children. We acknowledged that caregivers, especially parents, are their children's first teachers and recognized their expertise in adjusting activities to their children's capabilities. Compared to our staff, the caregivers know best how to present information to set their children up for success.

Therefore, the team focused on positioning caregivers as co-facilitators in the program and helping them understand the purpose behind the activities. We gave general instructions to the whole group but made sure to explain to parents the goals of each activity and offer some suggestions about how they could tackle it as a family. This small difference made a huge impact because it transformed caregivers from bystanders who shadowed and only occasionally helped their children into active participants who worked with their children on tasks like collecting and analyzing data. "Thanks to you . . . we are able to learn together," said one parent.

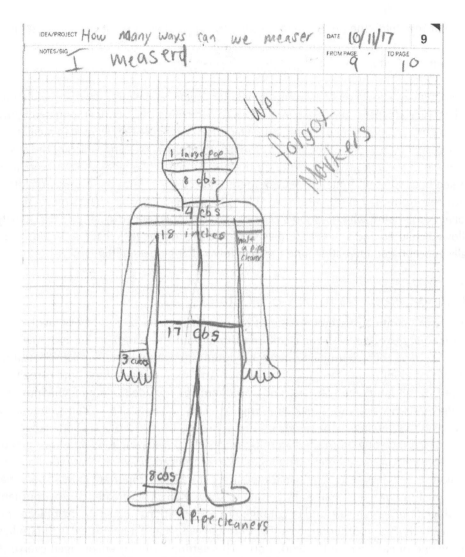

Figure 6.4 Data recorded by a child (this page) and her caregiver (next page) during the Measuring James activity.

Caregivers were empowered to adapt activities to children's interests and needs, discuss lesson plans with staff, interpret the activities in their own ways, and share in children's learning experiences. It also helped to create a community of caregivers within the program who helped each other – for example, by offering translations, support, or instruction – and who shared information and discussed topics related to their children's education, such as selecting their children's schools.

We also provided caregivers with materials that respected their roles as equal and necessary participants in the program. We gave them their own notebooks and measuring tools so they could record their own data (see Figure 6.4) and provided them with materials they could use to create their own final projects alongside their children.

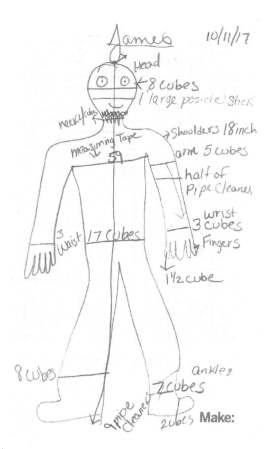

Figure 6.4 Continued

In addition, some activities were deliberately designed to prompt parent–child collaborations. For example, some activities required the work of more than one person. It took two people to measure the height of an exhibit that was too tall for children to reach or to measure how much time a participant played at an exhibit, with one person playing and another working a stopwatch (see Figure 6.5). Other activities, such as arranging or interpreting a large dataset, were designed to be conceptually challenging for children to complete by themselves. These strategies gave caregivers a specific role to play, in addition to offering a larger learning experience.

Supporting Caregivers' Needs

Putting caregivers in the role of facilitators often required additional preparation (Chandler-Campbell et al., 2020). Not only did caregivers need to learn about the topics we presented, but they also needed time and resources to take care of the whole family and accommodate siblings of different ages. Therefore, preparation and support needed to be timely, efficient, and effective. For instance, to help caregivers understand the broader context of the workshop activities, we provided them with a one-page handout that contained key information about data modeling and

Figure 6.5 A child and his caregiver working together to measure an exhibit.

suggested open-ended questions that caregivers could use to start conversations with their children. These open-ended questioning strategies helped facilitate deeper conversations and gave caregivers an accessible starting point for discussions about a potentially daunting topic. The handouts, which were available in English, Spanish, and Chinese, were particularly helpful for caregivers who had little experience in leading open-ended conversations about STEM with their children. This informa-tion boosted their confidence to engage in these types of conversations with their children and with each other. In addition, staff modeled similar prompts in their conversations with children.

We found that providing an overview of workshop goals and expectations and strategies for starting conversations without needing to "know the answers" allowed caregivers to more easily help children of different ages. They could jump in and out of tasks as needed while still monitoring and following their children's work and modify the activities for their younger children. We also frequently used familiar materials like measuring tapes, so that caregivers could focus on the larger ideas related to data without having to master a new tool themselves before they could help their children.

Conclusion

It is a challenging task to create inclusive and equitable programs that engage chil-dren and their families in learning about complex STEM topics like data modeling. In this research and development project, we generated new principles and strategies

c. james liu et al.

for addressing this issue. We iteratively developed and refined a family-focused workshop that engaged children and their caregivers in learning together through hands-on exploration and data-driven investigation. Overall, we believe that fostering families' agency is the key to supporting and engaging families from cultures and linguistic backgrounds that remain underserved in STEM fields. Specifically, our program focused on supporting caregivers' agency and inviting them to be active participants in our learner-centered approach to STEM education.

This work shows the value of having museum educators, researchers, and program developers work together to identify the needs of their target audiences and provide culturally and developmentally appropriate support. By conscientiously and thoughtfully understanding our practices and our visitors, science centers can help provide richer opportunities for families to create shared learning experiences that open up possibilities for deep and ongoing engagement with STEM.

Acknowledgments

The authors would like to thank Catherine Cramer, Jennifer Cumbe, Nadia Hajidin, Peggy Monahan, Madiha Naeem, Antonio Renovales, Laycca Umer, Stephen Uzzo, Janella Watson, and Anika Zaman for their contributions to the research, development, and implementation efforts throughout this project. We also thank Katy Börner, Maurice J. Elias, Herbert P. Ginsburg, Kim Kastens, Kathleen McKeown, Richard Lehrer, and Andee Rubin for providing guidance and suggestions throughout the project. This presented work was supported by the National Science Foundation (Grant No. 1614663). Any opinions, findings, and conclusions or recommendations expressed in this material are those of the authors and do not necessarily reflect the views of the National Science Foundation.

References

Ash, D. (2003). Dialogic inquiry in life science conversations of family groups in a museum. *Journal of Research in Science Teaching, 40*(2), 138–162.

Ben-Zvi, D., & Garfield, J. B. (2004). Statistical literacy, reasoning, and thinking: Goals, definitions, and challenges. In D. Ben-Zvi & J. B. Garfield (Eds.), *The challenge of developing statistical literacy, reasoning and thinking* (pp. 3–16). Kluwer Academic Publishers.

Callanan, M. A., Castañeda, C. L., Luce, M. R., & Martin, J. L. (2017). Family science talk in museums: Predicting children's engagement from variations in talk and activity. *Child Development, 88*(5), 1492–1504.

Chandler-Campbell, I. L., Leech, K. A., & Corriveau, K. H. (2020). Investigating science together: Inquiry-based training promotes scientific conversations in parent-child interactions. *Frontiers in Psychology, 11*, 1934.

Clements, D. H., & Sarama J. (2014, March 3). Play, mathematics, and false dichotomies. *Preschool Matters Today*. National Institute for Early Education Research. http://nieer.org/2014/03/03/play-mathematics-and-false-dichotomies

Crowley, K., & Jacobs, M. (2002). Building islands of expertise in everyday family activity. In G. Leinhardt, K. Crowley, & K. Knutson (Eds.), *Learning conversations in museums* (pp. 333–356). Lawrence Erlbaum Associates.

Dawson E. (2014) Equity in informal science education: Developing an access and equity framework for science museums and science centres. *Studies in Science Education, 50*(2), 209–247. https://doi.org/10.1080/03057267.2014.957558

Dawson, E. (2019). *Equity, exclusion and everyday science learning: The experiences of minoritised groups.* Routledge.

Donnelly, K., Liu, C. J., Meza, D., Letourneau, S. M., Umer, L., Uzzo, S., & Culp, K. M. (2018). *Museum makers: Designing with data.* Online curriculum. https://nysci.org/wp-content/uploads/Big-Data-for-little-Kids-sm.pdf

English, L. D. (2012). Data modeling with first-grade students. *Educational Studies in Mathematics, 81*(1), 15–30.

English, L. D., & Watters, J. J. (2005). Mathematical modelling in the early school years. *Mathematics Education Research Journal, 16*(3), 58–79.

Falk, J. H., & Dierking, L. D. (1992). *The museum experience.* Whalesback Books.

Falk, J. H., & Dierking, L. D. (2018). *Learning from museums.* Rowman & Littlefield.

Gopnik, A., Meltzoff, A., & Kuhl, P. (1999). *The scientist in the crib: Minds, brains and how children learn.* William Morrow.

Greenes, C., Ginsburg, H. P., & Balfanz, R. (2004). Big math for little kids. *Early Childhood Research Quarterly, 19*, 159–166.

Haden, C. A., Jant, E. A., Hoffman, P. C., Marcus, M., Geddes, J. R., & Gaskins, S. (2014). Supporting family conversations and children's STEM learning in a children's museum. *Early Childhood Research Quarterly, 29*(3), 333–344.

Jackson, K., & Cobb, P. (2010, April 29–May 4). *Refining a vision of ambitious mathematics instruction to address issues of equity* [Paper presentation]. American Educational Research Association, Denver, CO, United States.

Konold, C., Higgins, T., Russell, S. J., & Khalil, K. (2015). Data seen through different lenses. *Educational Studies in Mathematics, 88*(3), 305–325.

Leinhardt, G., Crowley, K., & Knutson, K. (Eds.). (2002). *Learning conversations in museums.* Lawrence Erlbaum Associates.

Lehrer, R., & Schauble, L. (2002). *Investigating real data in the classroom: Expanding children's understanding of math and science.* Teachers College Press.

Lehrer, R., & Schauble, L. (2004). Modeling natural variation through distribution. *American Educational Research Journal, 41*(3), 635–679.

Letourneau, S. M., Liu, C. J., Donnelly, K., Meza, D., Uzzo, S., & Culp, K. M. (2020). Museum makers: Family explorations of data science through making and exhibit design. *Curator: The Museum Journal, 63*(1), 131–145.

Lobo, P., & Salvo J. (2013). *The newest New Yorkers: Characteristics of the city's foreign-born population–2013 Edition.* New York City Department of Planning. www1.nyc.gov/assets/planning/download/pdf/planning-level/nyc-population/nny2013/ac.pdf

Makar, K., & Rubin, A. (2009). A framework for thinking about informal statistical inference. *Statistics Education Research Journal, 8*(1), 82–105.

McClain, L. R., & Zimmerman, H. T. (2014). Prior experiences shaping family science conversations at a nature center. *Science Education, 98*(6), 1009–1032.

National Research Council. (2009). *Learning science in informal environments: People, places, and pursuits.* The National Academies Press.

National Research Council. (2013). *Next Generation Science Standards: For states, by states.* The National Academies Press. https://doi.org/10.17226/18290

New York City Department of Health (NYCDOH). (2015). Community health profiles 2015: Queens Community District 4: Elmhurst and Corona. NYCDOH. www1.nyc.gov/assets/doh/downloads/pdf/data/2015chp-qn04.pdf

Pattison, S. A., & Dierking, L. D. (2018). Early childhood science interest development: Variation in interest patterns and parent-child interactions among low-income families. *Science Education, 103*(2), 362–388.

Pfannkuch, M. (2018). Reimagining curriculum approaches. In D. Ben-Zvi, K. Makar, & J. Garfield (Eds.), *International handbook of research in statistics education* (pp. 387–413). Springer.

Philip, T. M., Olivares-Pasillas, M. C., & Rocha, J. (2016). Becoming racially literate about data and data-literate about race: Data visualizations in the classroom as a site of racial-ideological micro-contestations. *Cognition and Instruction, 34*(4), 361–388.

Pianta, R. C., & Harbers, K. L. (1996). Observing mother and child behavior in a problem-solving situation at school entry: Relations with academic achievement. *Journal of School Psychology, 34*(3), 307–322.

Rogoff, B., Callanan, M., Gutiérrez, K. D., & Erickson, F. (2016). The organization of informal learning. *Review of Research in Education, 40*(1), 356–401.

Schoenfeld, A. H. (2002). Making mathematics work for all children: Issues of standards, testing, and equity. *Educational Researcher, 31*(1), 13–25.

Schulz, L. E., & Bonawitz, E. B. (2007). Serious fun: Preschoolers engage in more exploratory play when evidence is confounded. *Developmental Psychology, 43*(4), 1045–1050.

Siegel, D. R., Esterly, J., Callanan, M. A., Wright, R., & Navarro, R. (2007). Conversations about science across activities in Mexican-descent families. *International Journal of Science Education, 29*(12), 1447–1466.

Thomas, D. R. (2006). A general inductive approach for analyzing qualitative evaluation data. *American Journal of Evaluation, 27*(2), 237–246.

chapter 7

Designing Maker Programs for Family Engagement

David Wells, Susan M. Letourneau, and Samantha Tumolo

> [I liked] watching him use tools he's never used before . . . Now I know that you can teach kids about this kind of stuff, because I wouldn't have known how to do that.
>
> A caregiver participating with her child in a walk-up maker program

A long line of research has demonstrated the benefits of making – designing and constructing artifacts that their creators see as useful and intrinsically rewarding – for many aspects of children's learning in science, technology, engineering, and mathematics (STEM). Making and tinkering provide opportunities for children and youth to build content knowledge, practice skills relevant to STEM disciplines, and strengthen emerging interests in science and technology by tackling creative projects with support from communities of fellow learners (Vossoughi & Bevan, 2014). Recognizing that caregivers are a critical part of children's learning, museums and science centers have developed family-focused maker experiences that introduce the tools, materials, and ways of thinking involved in making and that support families in learning together (Brahms & Crowley, 2016; Brahms & Werner, 2013; Roque, 2016).

This chapter describes a research and development project at the New York Hall of Science (NYSCI) that explored how making and tinkering experiences can be designed to appeal to both children and their caregivers, particularly those with little prior experience. Through our development of making and tinkering programs for children and their caregivers from our local community, we investigated how the tools, materials, and facilitation strategies used in these experiences could support children's learning and caregivers' engagement as facilitators, observers, or makers themselves.

In this chapter, we describe our iterative process for designing a series of these making and tinkering programs. We share findings from our research – which included observations of more than 200 family groups and interviews with 88 caregivers – about which strategies were most effective in supporting caregivers' and children's engagement. To give a sense of how families engaged with the program, we also include case studies of the learning experiences of particular families. Based on evidence from the project, we make suggestions for designing and facilitating family-centered programs.

Program Context and Background

This research and development project was built on longstanding practices established in NYSCI's Maker Space for creating engaging programs (New York Hall of Science, 2013). Maker Space at NYSCI is a learning environment where children, teens, adults, and families can tinker, design, and create together. We reuse everyday materials like cardboard tubes, wire, and scraps of wood in exciting ways. We encourage experimentation and open-ended exploration and believe that making mistakes is a great way to learn.

Context

NYSCI is located in the Corona neighborhood of Queens, NY, one of the most diverse areas in the world. Our surrounding neighborhood is home to many first-generation immigrants from many parts of the world, particularly Central and South America. NYSCI has a long history of partnering with schools and community organizations in Corona to create rich STEM learning opportunities that respond to the needs and interests of local families. (See Chapter 5 for a more detailed discussion of NYSCI's community engagement work.) We provide individuals and families with a safe space to explore materials and use tools to support STEM learning. We offer a variety of different programming throughout the year, but in this chapter, we focus on the development of walk-up programming to support families' engagement.

Pedagogical Approach

In Maker Space, our educational approach accentuates the possibilities of tools and materials and stresses the importance of the process over the project outcome. Of course, we care about what visitors make in our programs, but we have found that if they focus on the skills necessary to create their projects, we see a strong sense of confidence emerge. This is revealed in the choices that learners make throughout the process – how they use tools to accomplish what they want or playfully explore tools to guide their creative process. Removing a specific project goal gives learners the opportunity to freely investigate a tool's potential and apply it to an intrinsically motivated project.

We start the process of designing new programming by asking ourselves questions like these: What tools and materials do we want visitors to explore? How might they use these tools and materials? What is our main objective? Answering these questions helps to ground the design process and keep us on track as we develop activities. We think deeply about play and how we can use playful entry points to invite our visitors into the space as well as the activity. We strive to create an inclusive learning environment that is as dynamic as our visitors' interests and experiences. We use observations and discussions with our visitors to guide this process and do a lot of testing and prototyping. We reflect on how to best engage visitors and facilitate their involvement in activities. We consider the efficiency of the room setup, how the materials and tools support the objectives, and how these factors shape the flow of the experience. Through reflection, we discover new possibilities for improvement.

Research Goals

After iteratively developing many previous Maker Space programs, we had already arrived at several effective strategies for engaging visitors in using tools and materials to realize their ideas. In this project, we were interested in finding out how our programs could better engage families, particularly caregivers who visited with their children. Decades of prior research have demonstrated the vital role that parents and caregivers play in children's learning experiences outside of school. Within museums and science centers, caregivers scaffold children's understanding of new ideas, connect to children's prior experiences and interests, and share in collaborative inquiry and exploration (Callanan et al., 2020; Crowley & Jacobs, 2002; Falk & Dierking, 2018; Zimmerman et al., 2010). In maker programs, caregivers are often involved in supporting children's learning, but they may need guidance and support themselves to participate in productive ways (Brahms & Werner, 2013).

Despite this research, we know relatively little about what draws family groups to maker programs, particularly families from non-dominant cultural and linguistic backgrounds. Because informal learning is grounded in families' prior experiences, cultural practices, and ways of knowing (Bang & Medin, 2010; Rogoff et al., 2016; Vossoughi et al., 2016), it is productive to consider what families value about these experiences and to provide opportunities for both children and caregivers to exercise agency in shaping their own learning experiences. Our approach in this project drew on sociocultural theories of informal learning in museums (Falk & Dierking, 2018; Rogoff et al., 2016). It was also informed by research in community psychology, which argues that physical and social settings jointly influence behavior at an individual, family, and community level (Gomez & Yoshikawa, 2017; Seidman & Capella, 2017).

Based on these perspectives, we expected that physical aspects of the environment and the materials used in the programs might affect whether and how families participated. We also surmised that facilitation and opportunities for social interaction would be crucial in inviting adults to explore, tinker, play, and make alongside their children. We investigated these hypotheses by developing two distinct maker

david wells et al.

programs that took place in different physical spaces. Both programs used a range of tools and materials, but they were facilitated in slightly different ways. This allowed us to observe which qualities of the programs shaped families' participation as a whole as well as caregivers' involvement.

Designing the Programs

Our plan was to develop a series of self-directed, tabletop, exploratory making experiences in our Science Library and another series of walk-up activities for children and adults in Maker Space. To differentiate these experiences, we focused on the tools and materials we would use in each setting, the concepts we wanted visitors to explore, and the distinct facilitation styles that best supported each type of experience.

Tabletop Activities

The tabletop activities in the Science Library would use familiar materials in interesting ways. They would require a low level of facilitation, and facilitators would take a lighter touch in providing guidance on tool use. The activities were meant to support brief engagements, while allowing visitors to explore a concept more deeply if they were so inclined. Creating engaging entry points for these activities was critical because they were situated on a table and had to appeal to passersby. We knew from past experience that using everyday materials in interesting ways caught people's attention.

An activity called Marble Run illustrates this approach. In Marble Run, visitors used an array of materials – cardboard tubes, kitchen utensils, bells, gears, strawberry baskets, and other random objects we thought would provide inspiration – to create a pathway on a wall. They would test their design by dropping a marble and watching it roll down through all the components until it hit a bell or landed in a cup, whichever the designer decided the finale would be. This activity called for kinesthetic engagement – standing up, moving around, and working feverishly to secure components, test, iterate, and try again. This process was often accompanied by cheers, clapping, and oohs and aahs, which further attracted visitors to watch and participate. This kind of ideal entry point instantly engages children and adults and has potential for sustained and collaborative engagement. It invites engagement on two levels: the use of everyday materials in unexpected ways draws visitors closer, and the physical and social nature of the activity sustains engagement and attracts other visitors. This activity also enables multiple people to participate at the same time, so adults and children can work together or side by side. We designed and prototyped a variety of tabletop maker activities that use physical and social engagement to inspire this exploration.

Maker Space Walk-Up Activities

When designing walk-up activities for Maker Space, we put the central focus on a diverse set of tools and materials within the space. We planned to have more facilitation for these activities than for the tabletop activities, to allow families to explore

Figure 7.1 Children learn how to use a scroll saw in Maker Space.

a wider variety of tools. These included hammers, saws, and power tools, as well as digital tools, textiles, and molding and casting tools (see Figure 7.1). Materials included wood, foam insulation, nails, fabric, rubber bands, hot glue, and many others, depending on the tools being offered.

We planned to use these tools and materials to accomplish multiple goals. We hoped that each visitor would find something they connected with and could discover a new interest or passion. Ideally, we also wanted to provide a meeting space where families felt comfortable and confident in approaching new experiences and sharing their own ideas and perspectives. One way we accomplished this was by inviting visitors to contribute to a collaborative installation after learning how to use tools in Maker Space. These collaborative art pieces evolved over time as multiple groups of visitors added their work. This helped to give the visitors a sense of ownership and belonging in Maker Space – a "third space" between home and school that they could call their own. We noticed that visitors often returned to Maker Space for multiple activities and expressed a connection to the collaborative projects as things they had worked on.

As we continued to design the series of tool-based walk-up activities, we pulled from our Maker Space "greatest hits" – well-tested activities that have been implemented for several years and have consistently engaged family groups, so we knew they would work well in this setting. One such activity was Bucket of Junk, in which visitors use a hot glue gun to combine random materials that we had collected through donations and from our museum storage spaces. Building on this idea, we created a similar experience called Scrap-shop. We collected all the wood scraps we could find, filled

david wells et al.

up buckets with the scraps, and asked visitors to build anything they wanted. This created the opportunity to add a hammer and nails to the tools we were already using in Bucket of Junk.

In addition to reimagining existing activities, we developed and tested new activities. These included a virtual reality drawing activity, in which visitors created a drawing or collage and then used virtual reality goggles and the Panoform tool[1] to step into a 360-degree virtual image of their creation. Another example was an activity called Wires and Pliers. We provided 16-gauge wire, wire cutters, and pliers, and asked visitors to bend, cut, and shape the wire into anything they wanted. We also designed a shadow play area so visitors could hang their sculptures in front of a light and see what the shadows looked like.

For each of the tabletop and walk-up activities, we built and tested prototypes with a team of people in Maker Space. Then we tested the activity with small groups of visitors to see how it played out "in the field." This testing provided us with valuable information about which facilitation strategies were helpful for different types of learners and how to set up the space to be most conducive for family participation. In Maker Space, it is standard practice to reflect after each program with all the staff and to document and use this reflection to improve the experience. Based on the iterative revisions we made to the programs, we developed activity overviews and facilitation guides. These resources helped facilitators learn for themselves about the tools and materials, set up the programs, introduce and facilitate activities with visitors, and discuss tool safety concerns using positive language. These living documents served as the basis for ongoing training with facilitators and were revised and updated as we reflected together about the best ways of supporting children and their families.

Implementation and Findings

As part of NYSCI's community engagement initiative, this program has access to families from 15 local partner schools who visited NYSCI for free during after school hours and participated in a range of special programming and access to NYSCI's exhibits during those hours. Programming included cultural and other events, homework help, as well as the making and tinkering programs that we developed in this project, which provided an opportunity for participating families to learn how to use tools in our Maker Space.

A key goal for Maker Space programming developed for families was to foster a sense of ownership and belonging as STEM learners – for families to feel like Maker Space is "theirs" during their time at the museum and to use it to pursue their own ideas and interests. Because families participating in the NYSCI's community engagement initiative visit the museum regularly, their children have extended opportunities to build skills and confidence with a variety of tools and materials over time, and caregivers are invited to be a part of this process.

As we iteratively developed and revised the programs over the course of a year, we sought to determine which aspects of the programs made a difference in supporting families' engagement. Each week, the team discussed how families had participated and how the programs could be improved. We also gathered field notes about how more than 200 family groups engaged with the programs – including the roles that caregivers played in supporting children's learning and the aspects of the environment that influenced how they were involved. Finally, we interviewed 88 caregivers who participated in the programs. We asked them to describe what they felt their children were learning from the programs, how they were involved, what they valued about these experiences, and how the programs related to their previous experiences with making or other STEM activities at home. Through discussions and analysis of these observations and interviews, we homed in on aspects of the programs that supported the overall participation of families, as well as the active involvement of children and caregivers.

Inviting Families to Participate

Early in the implementation of the programs, we developed strategies to tailor the programs to families' needs. For example, children typically completed their homework before attending other programs and would attend on days when they had time. We usually saw families once or twice a week, but family members often participated at different times or on different days. To make sure each person would be able to experience each activity or tool if they wanted to, we offered the same program activity in week-long intervals. To keep things from being too repetitive for those who participated more than once a week, we offered slight variations each day. Children would sometimes come back the next day with a new idea for a particular project, allowing us to see increases in both their skill level and confidence.

In addition, we noticed that the location of the activities had a major impact on whether families chose to participate. Accordingly, we moved programs to different locations to make them more inviting. For example, many visitors were hesitant to enter Maker Space, which is located in a separate area away from the rest of the after-school programs and museum exhibits. Families tended to remain in parts of the museum where they could split up and still remain visible to one another and to revisit familiar areas that they knew children of different ages would enjoy. We decided to move the programs to a more central location in an open area called the Sandbox that was clearly visible from elsewhere in the museum. This made the space more accessible and welcoming, and closer to other existing family programming. We ran programming in the Sandbox for a few weeks to allow families to become more familiar with our staff and the types of programs we offered. Once this trust was established, we moved the programs back into Maker Space.

Similarly, the tabletop activities were moved from the Science Library, which was in a closed-off area of the museum far from the homework help program, to pop-up tinkering stations between exhibits and near the cafeteria, where families tended to congregate. A Maker Space resident would engage families in the tinkering activities and work to establish a rapport with families with whom we had not yet interacted. Once this rapport had been established, the resident staff would show families examples of the projects happening in Maker Space that week and encourage them to try the tool-based programs for a more in-depth maker experience. By strategically placing programs in different areas of the museum and working to build trust and familiarity with our staff and our programming, we were able to invite families to try activities that they might have overlooked.

Understanding Caregivers' Roles

Throughout the development and implementation of the programs, we endeavored to understand the range of ways in which caregivers might participate and their perceptions of the value of these experiences for their children and themselves. Across the 88 caregivers that we interviewed, 38% were primarily involved in facilitating their children's learning during the programs (based on our observations and caregivers' own descriptions of their involvement). In interviews, these caregivers spoke about wanting to spend time with their children and described learning from facilitators about activities they could do at home or ways to help their children with hands-on projects. Another 30% of caregivers mainly observed what children were doing, either from a distance or from nearby. These caregivers emphasized children's independence and described how they noticed children's interests and abilities while watching them approach a new learning experience, particularly those involving tools that children had never used before. Finally, 21% of caregivers directly participated in the activities as makers. They described the process of creating something by hand as the most beneficial part of the experience for their children and themselves. They often mentioned that these experiences had tapped into their creativity and made them feel they had "permission" to play and be creative.

These findings show that caregivers were engaged in a variety of ways – there was no single "best" or most effective type of engagement. In fact, caregivers described unique benefits of different types of involvement. Based on these findings, we focused on creating space for a range of caregiver–child interactions and invited caregivers to observe, facilitate, or participate depending on their preferences and family dynamics.

Designing to Support Children and Caregivers Together

Through our discussions and ongoing analysis, we identified specific aspects of the programs that shaped how children participated and how caregivers were involved.

Room Set Up and Materials Distribution

We experimented with different ways to set up the space to allow visitors to safely focus on what they were making while forging a sense of community. Individual workstations enabled children to focus with fewer distractions and gave them ample space in which to work, but it did not lead to as much collaboration or connection with other visitors. Moreover, using individual workstations seemed to convey that the activities were meant only for children; caregivers tended to be involved as observers or facilitators, rather than makers. Still, caregivers reported that they appreciated having dedicated spaces in which children could focus on what they were doing.

As an alternative, we pushed tables together to create a communal workspace where everybody was creating together. This encouraged families to sit and stay for long periods of time, as they collaborated, shared stories, and developed connections with each other and our staff. The communal workspace also supported caregivers' involvement as facilitators and as makers themselves. While this setup worked for smaller scale projects, such as sewing and felting, it was not conducive to all activities. For example, woodworking required more space and separation from others.

In short, there were pros and cons to both the individual and communal arrangements that had to be considered in light of the goals of the particular activity. For each activity, we found a balance between the safety considerations and the communal setting and family engagement we wanted to establish.

In addition to room setup, the distribution of materials played an important role. For many activities, we preferred to place materials in reachable containers ("family style") so visitors could take what they needed when they needed it. However, this did not work for all activities. Sometimes children became so excited by the array of materials that they took as much as they could, and their projects lost focus or we ran out of materials before others were able to create anything. We often set up a separate materials station, where we encouraged children to tell us what they needed and what their plan was before beginning to build. This helped to keep their projects more organized and intentional and conserve resources. It also allowed caregivers to be involved in each step of the process.

Access to Novel Tools and Materials

Many families said that the programs were novel to them because they did not have access to the same tools, materials, and expertise elsewhere. The ability to experiment and try out new tools from week to week was a major draw for participating families. We found that novel or high-tech tools and materials, such as virtual reality goggles, 3D pens, and scroll saws, motivated caregivers to engage directly in making. Many caregivers described acquiring new skills as a key benefit for their own learning and well-being.

In contrast, more familiar and everyday tools and materials, such as fabric and hand tools, were effective in supporting caregivers as facilitators of their children's learning. Caregivers were more likely to link activities that used familiar tools and materials to their families' other interests or experiences at home or in school. For our team, this highlighted the value of including both novel and familiar items to tap into families' existing skills and knowledge while enabling families to experience something new together.

Family-Centered Facilitation

We encouraged facilitators to use an inquiry-based approach that involved asking a lot of questions to give them a chance to listen to and understand visitors' perspectives. Using questions as a starting point for a conversation and inviting visitors to take a moment to think before responding kept children and their caregivers engaged and present. It also helped facilitators consider how they could assist families in getting started or working through challenges together. Additionally, we made real-world connections by encouraging families to think about where they had seen these tools or materials before and to talk about what they already knew about how each tool was used. Our goal was to demystify each tool and explore the associations families might have with them, in order to encourage them to see any tool or material as an open-ended pathway to creativity or problem-solving.

We also adjusted how we addressed families in order to include caregivers in the process. We found that when facilitators gave instructions and tips to *caregivers*, the caregivers tended to take on the role of co-facilitator, passing along this knowledge to children and providing assistance. When facilitators spoke directly to *children*, caregivers were more likely to defer to the facilitator and observe. We also found that allowing all members of the group to immediately engage with the materials communicated that adults were also welcome to make and tinker with their children. Based on these observations, facilitators developed a range of strategies to be responsive to families' dynamics. These included giving brief introductions and inviting everyone to explore the materials, stepping back in order to allow caregivers to be the primary facilitators or to try using tools and materials themselves, and stepping in to help children when caregivers were engrossed in their own projects.

Case Studies of Children and Caregivers as Learners

As children became more familiar with the museum in general and Maker Space in particular, they began expressing greater ownership over their projects and using the space as a freely available resource. As the program progressed, we also saw growth in children's agency. For example, as children gained skills and confidence in Maker Space, they sometimes worked completely independently. At times, children were the ones teaching caregivers how to use the tools or suggesting new project ideas to museum staff. These long-term learning experiences empowered both children and caregivers. Children took greater control of decisions about what they wanted to learn, and caregivers took greater advantage of opportunities to explore tools and materials themselves and notice the skills and confidence that children were developing. The cases below illustrate the diverse ways in which the programs supported children and caregivers as learners.

K and Her Family

Most families that regularly attended the after-school program at NYSCI came to Maker Space once or twice a week, depending on how much homework children had. Two families came every day, often multiple times a day, and sometimes when Maker Space wasn't even open. One family included two children, approximately 10 and 8 years old. The older child, K, was shy at first but had a natural interest and talent in the arts. K always came to Maker Space to work on whatever activity was being offered. She would stay for long periods to make her projects as detailed and close to her vision as possible. Even if her mother and brother came into the space, she typically worked independently.

As K became more comfortable with the staff and familiar with the tools, materials, and activities offered, she began to assume greater agency in using the space. She began by asking if we had additional supplies or materials that she could use to improve her projects, which often led to other visitors wanting those supplies as well. This helped us improve our program design because we learned more about what materials and aesthetics appealed to our younger visitors. Additionally, this demonstrated to K, albeit on a small scale, that her voice mattered and that this was a place where she could express agency and create change.

K's sense of agency and belonging within the space continued to evolve throughout the year. If other staff members were working on projects for different programs, she often would approach them and either ask to help or ask if they could teach her what they were doing. As a result, she developed more niche skills like woodcarving. There were several weeks when most visitors were working on a relatively simple project, and K was happily working on an elaborate wood carving in the back of the room. She often chose to work on projects that differed from those we were focusing on with other visitors, but to us, this symbolized that we were successfully creating a space where visitors felt they had a voice and creative outlet. Toward the end of the semester, K began suggesting projects for us to teach. She described step by step how we would go about making her idea come to life, essentially mimicking how we typically introduced a project to visitors. This behavior confirmed that we had succeeded in creating an inclusive space where she was able to take ownership over her experience.

As this progression took place, our staff also developed a relationship with K's mother, who always stopped to say hi and see what K was working on even if she wasn't participating in that day's activity. Although her mother's English was limited, K translated for her, and we were able to share jokes and stories as we made things together. Several families, including K's, gave the staff nicknames, and some even gave staff presents on holidays, demonstrating to us that caregivers valued the family-friendly atmosphere in the space. K's mother seemed to recognize that this space not only provided opportunities for K to express herself, but also welcomed her as a parent to observe her daughter engaging in the activities.

C's Experience

Another family that attended regularly included a mother and C, her seven-year-old daughter. C typically came to Maker Space at the beginning of the after-school session to see what the activity each day was about. If the activity focused on something new to her, she would quickly learn how to use the tool and get started on her project before running back upstairs to finish her homework. During the remainder of the session, she would come back periodically to continue working on her project, often bringing other children with her. She would introduce us to whomever she brought, before teaching them how to use the tool and getting started on the project herself. In short, she exercised agency over the space that reflected her confidence with both using the tools and teaching others. She watched how we would introduce a tool or activity and felt enough of a sense of ownership over the space to follow this same process to help newcomers. Because of her enthusiasm to teach others, she often wouldn't have time to finish her own project. She would ask us to hold onto it for her, reflecting her comfort with staff and with the space as a whole.

J and His Mother

Even within individual program activities, families exercised agency by seeking out new experiences together. For example, one family (a mother and her six-year-old son, J) participated in a tool-based activity on a weekday afternoon. Scroll saws – small electric saws that are used to cut intricate designs in cardboard, wood, metal, or plastic – were set up in the Sandbox area of the museum, an open space clearly visible from the exhibits on the museum's upper level. This family had seen the activity happening from above and, after watching other groups using the tool, had decided to come downstairs and give it a try. We gave J and a boy from another family group a brief explanation about how to use the scroll saw, as J's mother watched nervously from nearby. While the two boys worked independently, chopping up rectangles of cardboard into smaller pieces to become familiar with the tool, we shared some information with J's mother. We explained the safety features of the tool and described which materials were easier or harder to cut.

J interrupted to ask his mother to help him by drawing an outline of a train on a piece of cardboard, so he could cut it out. While the mother drew, J asked the other boy what his name was, and they introduced themselves. J's mother watched, listened, and laughed as the boys figured out how to share the materials on the table. She explained, "They're both six years old, you see. They think the same way." She watched closely and offered some suggestions, letting J take the lead. Later, J began cutting an outline of the letters in his name. He showed us what he did and asked a question about how to make one of the letters. J's mother said, "Now that's the evolution of his thinking right there. He started out making lines and now he's made his whole name. That's what he's gotten out of it, the confidence to know how to do that."

J's mother explained that during the program she had learned by watching how facilitators had helped her son use tools safely and how he had responded to this novel experience:

> It helps me loosen the reins a little bit . . . It was a good exercise to allow him to do it, even though it was hard. He watches his dad use tools, but he's never used them himself. It just shows that he can take risks in a safe manner with some supervision.

J's mother was actively involved throughout the process in her own way – she helped J with his project, provided encouragement, noticed his progress, and made connections to their lives at home. When J finished cutting out his name, a Maker Space staff member asked if she could take a picture of his project. He smiled broadly while she took a photo. J's mother asked him whether he was ready to leave, and he said, "No, I want to wait for my friend." When the other boy was finished, they proudly showed each other what they had made, and then each went their separate ways, taking their projects home with them. Even in this brief encounter, the setup and facilitation of the activity allowed everyone to feel welcome and exert agency in shaping their own learning experience.

Conclusion

By comparing a wide range of different styles of programming over a short period of time, this project allowed us to test and refine strategies that had been developed through many years in the museum's Maker Space and to directly observe the impact on families' engagement. We found that a variety of strategies could create a welcoming space for children and their caregivers to participate in maker programs and that the physical setup and facilitation style could be adjusted to invite caregivers to be actively involved.

These findings suggest ways to make programs more inclusive for families with a range of configurations, backgrounds, and interests. They point to the importance of opening up multiple ways for families to participate, rather than suggesting that there is one "right" way for caregivers to learn with their children. The approaches used by facilitators – to welcome caregivers into the process and offer support in light-touch and open-ended ways – helped to communicate that the experience was available to all. This allowed families to choose how to interact with the programs, with staff, and with one another in ways that made sense for them.

Acknowledgments

This material is based on work supported by the National Science Foundation under Grant No. 1723640. Any opinions, findings, and conclusions or recommendations expressed in this chapter are those of the authors and do not necessarily reflect the views of the National Science Foundation. The authors would like to thank additional Maker Space staff who were involved in this project, including Annalise Phillips, Danny Kirk, and Cesar Villar, as well as the families who participated in this project. We also thank project advisors Jill Denner, Paula Hooper, Vera Michalchik, and Anne Sekula.

Note

1 https://panoform.com/.

References

Bang, M., & Medin, D. (2010). Cultural processes in science education: Supporting the navigation of multiple epistemologies. *Science Education, 94*(6), 1008–1026.

Brahms, L., & Crowley, K. (2016). Learning to make in the museum: The role of maker educators. In K. Peppler, Y. Kafai, & E. Halverson (Eds.) *Makeology: Makerspaces as learning environments* (Vol. 1, pp. 15–29). Routledge.

Brahms, L., & Werner, J. (2013). Designing makerspaces for family learning in museums and science centers. In M. Honey & D. E. Kanter (Eds.), *Design, make, play: Growing the next generation of STEM innovators* (pp. 89–112). Routledge.

Callanan, M. A., Legare, C. H., Sobel, D. M., Jaeger, G. J., Letourneau, S., McHugh, S. R., Willard, A., Brinkman, A., Finiasz, Z., Rubio, E., & Barnett, A. (2020). Exploration, explanation, and parent–child interaction in museums. *Monographs of the Society for Research in Child Development, 85*(1), 7–137.

Crowley, K., & Jacobs, M. (2002). Building islands of expertise in everyday family activity. In G. Leinhardt, K. Crowley, & K. Knutson (Eds.), *Learning conversations in museums* (pp. 333–356). Lawrence Erlbaum Associates.

Falk, J. H., & Dierking, L. D. (2018). *Learning from museums.* Rowman & Littlefield.

Gomez, C. J., & Yoshikawa, H. (2017). Approach to structure and culture in family interventions. In M.A. Bond, I. Serrano-García, & C.B. Keys (Eds.), *APA handbook of community psychology: Vol. 1. Theoretical foundations, core concepts, and emerging challenges* (pp. 337–352). American Psychological Association.

New York Hall of Science (2013). *A blueprint: Maker programs for youth.* https://nysci.org/wp-content/uploads/nysci_maker_blueprint.pdf

Panoform. https://panoform.com/

Rogoff, B., Callanan, M., Gutiérrez, K. D., & Erickson, F. (2016). The organization of informal learning. *Review of Research in Education, 40*(1), 356–401. https://doi.org/10.3102/0091732X16680994

Roque, R. (2016). Family creative learning. In K. Peppler, Y. Kafai, & E. Halverson (Eds.), *Makeology: Makerspaces as learning environments* (Vol. 1, pp. 47–63). Routledge.

Seidman, E., & Cappella, E. (2017). Social settings as loci of intervention. In M.A. Bond, I. Serrano-García, & C.B. Keys (Eds.), *APA handbook of community psychology: Vol. 2. Methods for community research and action for diverse groups and issues* (pp. 235–254). https://doi.org/10.1037/14954-014

Vossoughi, S., & Bevan, B. (2014). *Making and tinkering: A review of the literature.* Commissioned paper for Successful Out-of-School STEM Learning: A Consensus Study, Board on Science Education, National Research Council. https://sites.nationalacademies.org/cs/groups/dbassesite/documents/webpage/dbasse_089888.pdf

Vossoughi, S., Hooper, P. K., & Escudé, M. (2016). Making through the lens of culture and power: Toward transformative visions for educational equity. *Harvard Educational Review, 86*(2), 206–232. https://doi.org/10.17763/0017-8055.86.2.206

Zimmerman, H. T., Reeve, S., & Bell, P. (2010). Family sense-making practices in science center conversations. *Science Education, 94*(3), 478–505. https://doi.org/10.1002/sce.20374

chapter 8

Innovation Institute
Follow the Youth

David Wells, Elham Beheshti, and Danny Kirk

> There's more autonomy in [the Innovation Institute] ... Usually, programs tell you what you're going to learn and that you're going to do this specific project, but this is structured in a way that teaches you things, and you decide what you want to do with it.
>
> Student participant in the Innovation Institute

Rapid advances in technology and communications have dramatically affected education by changing how people learn, gather information, and engage with others. These advances also have the potential to expand pathways for youth into science, technology, engineering, and mathematics (STEM) careers. There are well-documented hurdles, however, to attracting and retaining underrepresented youth in STEM disciplines. Research on adolescents' decision making about their educational and career paths (Eccles, 2009; Wigfield & Eccles, 2000) and work on the design of effective STEM pipeline programs (Bailey, 2015; Ellis, 2015) suggest that even high-quality education programs may not be enough to inspire youth to pursue STEM-oriented pathways. Underrepresented youth, in particular, who participate in programs designed to prepare them for STEM careers may perceive important next steps in the STEM pipeline as too risky or irrelevant to their current lives and have trouble seeing themselves in STEM careers (Lauermann et al., 2015; Wigfield & Eccles, 2000).

Researchers are demonstrating that youth, particularly those from groups underrepresented in STEM careers, need both access to learning opportunities and support to help them develop positive social and collective identities as potential STEM professionals (Eccles & Wang, 2015; Wang et al., 2013; Wigfield & Cambria,

2010). Eccles (2009) defines "social/collective identity" as "those personally valued parts of the self that serve to strengthen one's ties to highly valued social groups and relationships – those aspects of their identities that tie them to the communities they value highly." Building these identities involves changing students' perceptions of STEM careers and helping them make connections between their own personal values and priorities and STEM career pathways.

To address these challenges, the New York Hall of Science (NYSCI) developed and tested the Innovation Institute (I2), which is centered on the idea that STEM learning can be accessed by anyone. I2 provides opportunities for high school students, including many from underserved groups, to engage with the world around them, find personal connections to STEM, express their voices, and take action. Participating students use thinking and making skills to identify community problems that interest them and work through the entire process of developing a product or program to address this problem. Program activities range from prototyping and creating a digital version of a game board to using computer-assisted design tools to program circuits, as well as taking observational walks through the local community and interviewing community members. In addition to building their technical skills, these students are developing a sense of agency, gaining insights into their neighborhoods, and acquiring skills such as collaboration and communication.

This chapter explains the core components of the theoretical model underlying I2. We describe the evolution of the I2 program over the past six years and the implementation of its most recent iteration in 2019–2020. We also share a case study of a participant to illustrate how the program unfolded last year and discuss preliminary findings from research we conducted on the current version of the program.

Background and Theoretical Framework

The practices used in the I2 program are based on a theoretical model with three connected components, each of which is explained in more detail in this section:

- *Social entrepreneurship,* which encourages young people to see themselves as active agents of social change through community-based problem-solving
- *Design thinking,* which engages youth in a creative process for exploring the sources and scope of a problem and devising a solution that emphasizes the user's needs
- *Computational making,* in which youth apply "computational thinking" approaches from engineering and computer science to design and make things they see as useful and meaningful

Social Entrepreneurship as a Context for Youth Development

Social entrepreneurship can motivate young people to pursue STEM academic work and careers by creating opportunities for them to build connections between STEM careers and social impact. Although the phrase "social entrepreneurship" is often used loosely to refer to any kind of socially relevant service or product, Martin & Osberg's definition (2007) has three key components: (a) it responds to a need in society to address an inequity or source of exclusion or oppression of a social group, (b) it is based on a "social value proposition" in which a product or service can respond to that inequity, and (c) it has an impact on society that shifts perspective and shows how similar change could occur in other settings if the product or service were brought to scale.

Our I2 model uses social entrepreneurship to help encourage young people to see themselves as active agents of social change by impacting their communities. The model supports youth as they move through a "needs sensing" process to identify a challenge in their community that they can address through design thinking and computational making.

Design Thinking as a Context for Community Problem-Solving

In our model, design thinking is a social and cultural practice "rooted in community-based forms of surviving and thriving" (Vossoughi et al., 2016). Design thinking includes such practices as inquiring, repurposing materials, sharing resources, and communicating (Wardrip & Brahms, 2015). Key to these practices is the process of problem scoping, in which an investigator explores the sources and reach of a particular problem. Problem scoping offers a critical opportunity for youth to reframe a problem in ways that are relevant to them. Through problem scoping, youth may discover that problems that appear to be technical or isolated are often socially situated, historically rooted, and politically complex (Kilgore et al., 2007).

Computational Making as a Context for Creating Solutions

The concept of "making" has taken on new energy with the growth of the maker movement, an international effort to create spaces, resources, and methods for sharing expertise and supporting people as they design and construct artifacts that are useful or meaningful in their lives. Although making often focuses on analog or digital fabrication tools, ranging from lathes to 3D printers, it may also draw on crafting tools and techniques, such as cooking, sewing, and textile production. In all of these areas, making emphasizes skill, creativity, and innovation. Making can be educationally meaningful because participants learn by designing and developing artifacts of personal significance to them (Kafai, 2006; Papert, 1993; Vossoughi & Bevan, 2014).

All kinds of making require problem-solving. In the I2 program, we encouraged participants to use a problem-solving approach known as computational thinking (National Research Council, 2010; Weintrop et al., 2016; Wing, 2008). Computational thinking uses practices that are a regular part of a wide range of engineering and computer science careers (Boy, 2013; Buechley et al., 2008; College Board, 2015; Weintrop et al., 2016). These practices include breaking down problems into simpler components (decomposition), recognizing patterns, removing unnecessary information to focus on what's important (abstraction), and developing step-by-step instructions (algorithms) for solving the problem and addressing errors. (See Chapters 11 and 12 for a more detailed discussion of computational thinking.)

A growing number of researchers and practitioners are using the term "computational making" to describe the intersection of computational thinking and making practices. Computational making can be a powerful strategy for not only attracting underrepresented students to computing fields (Benda et al., 2012; Rode et al., 2015) but also transforming how and what people learn across STEM disciplines. The introduction of new technological tools and creative processes in STEM learning has been shown to broaden STEM participation, particularly among women (Buechley et al., 2013; Peppler, 2013; Qiu et al., 2013), and to improve STEM learning outcomes (Peppler & Glosson, 2013; Peppler, 2013).

Drawing from this prior research, the I2 model posits that when youth have the opportunity to deeply explore problems in their local communities that are personally relevant and to use computational thinking approaches to respond to these problems, their perceptions of engineering, computer science, and other STEM disciplines will positively shift (Eccles & Wang, 2015) (see Figure 8.1). We contend that combining social entrepreneurship with design thinking and computational making has the potential to deepen students' engagement with computational tools and positively affect students' self-efficacy – their perceptions of their own capabilities to succeed.

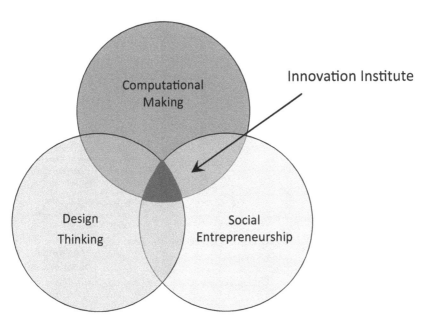

Figure 8.1 Our theoretical model is situated at the intersection of three core elements: computational making, design thinking, and social entrepreneurship.

Program Design

The current version of the I2 program is the product of four iterations of program implementation from 2014 through 2018. Drawing on lessons learned during this process, we adapted the design of the program in three important ways:

- We shortened the program from 12–15 months to 9 months, which our experience suggests is a feasible time frame for maintaining youth engagement.
- We lowered the target age group of the program from students in Grades 10–12 to those in Grades 9 and 10. This was done in response to research suggesting that early adolescents are particularly open to changes in their attitudes about academic paths and careers they may want to pursue (Wigfield & Eccles, 2000).
- We revised the program to sharpen the focus on computational approaches to problem-solving.

NYSCI also exposes I2 participants to STEM career pathways by partnering with outside organizations, makers, and entrepreneurs. These partners range from design thinking organizations like the global design company IDEO to start-ups and entrepreneurial makers who have carved out careers that address identified needs in their areas of interest.

The program starts in August and continues throughout the school year. Participants meet once a week for 2.5 hours after school and for a full day (10 a.m. to 4 p.m.) on the first Saturday of every month. The program culminates with the participants presenting their final projects and program experiences to an audience of peers, educators, and design industry professionals.

The I2 program consists of three main phases: launch, exploration, and application (see Figure 8.2). Each of these phases, described below, has been developed to create an emergent learning experience, meaning that youth are active participants in shaping the program as it unfolds.

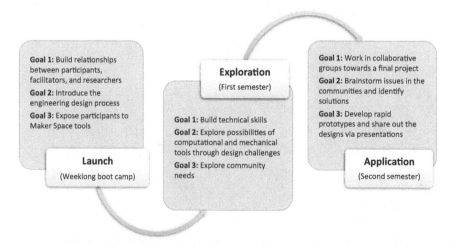

Exploration
(First semester)

Goal 1: Build relationships between participants, facilitators, and researchers
Goal 2: Introduce the engineering design process
Goal 3: Expose participants to Maker Space tools

Launch
(Weeklong boot camp)

Goal 1: Build technical skills
Goal 2: Explore possibilities of computational and mechanical tools through design challenges
Goal 3: Explore community needs

Goal 1: Work in collaborative groups towards a final project
Goal 2: Brainstorm issues in the communities and identify solutions
Goal 3: Develop rapid prototypes and share out the designs via presentations

Application
(Second semester)

Figure 8.2 The program flow.

david wells et al.

Phase 1: Launch

The launch phase starts with a weeklong boot camp in August and extends through the first half of the fall semester. Throughout this phase, we introduce the design thinking process developed by our project partner, IDEO, and involve students in group activities that encourage collaborative engagement. During the boot camp, we explore computational thinking concepts that will be used throughout the program. We also use design thinking and computational making to introduce rapid prototyping (techniques to quickly fabricate a model using everyday materials such as cardboard and tape, as well as 3D CAD software). We expose the participants to tools in our museum's Maker Space environment and discuss how to apply them to reach their objectives. This project-based approach has worked well in earlier versions of I2 to build a strong foundation of understanding materials, tools, and processes and being able to apply them.

Phase 2: Exploration

Once this foundation has been established, we shift the focus to building technical skills. We do this by engaging students in design challenges and prototyping sessions in which they explore community needs, explore the functionality of various tools, and research how those tools may be used to solve different problems. We have found that learning to use tools builds confidence and creative problem-solving skills. Through this process the group develops collective efficacy, which Bandura (1997) defines as a group's shared belief in its capabilities to organize and execute action for a desired outcome (p. 477). Students also have an opportunity to acknowledge and share their strengths with the group. The design challenges and prototyping sessions focus on building students' computational making skills; soft skills such as collaboration, communication, reflection, and presentation; and foundational skills in using tools.

In this exploration phase, students and NYSCI staff also participate in community walks to learn more about the neighborhoods in Corona, Queens, that surround NYSCI. The walks focus on community assets, such as stores, places of worship, parks, and other gathering places. Students observe how people engage with these places and spaces. They learn how to question their assumptions, become more socially aware, and make meaningful observations. I2 participants interview community members to hear about what these residents value in their community and what challenges they perceive. This helps participants to better understand the community and begin problem scoping.

During the problem-scoping process, participants collectively compile a list of overarching problem areas in response to the data collected through observations and interviews. Each participant makes a list of the problem areas that interest them the most. We use these lists to group students according to their interests. The groups then consider questions such as these: What is at the root of this problem? Are we able to solve it? How might we solve it? Asking these questions helps each group frame a process for determining which aspect of the problem they will focus on and compiling possible solutions. Through this process, program facilitators help participants to see connections between the skills they are developing and the computational tools that are available and to identify tools that would be beneficial for their particular projects.

Phase 3: Application

The third phase of the program emphasizes community engagement and problem-solving. Students decide on and implement a final project over eight to ten weeks that builds on what they learned in the first two phases. Working in small groups, students go through the full design process: generating ideas about a problem, selecting a problem to work on, using their available skills and tools, developing rapid prototypes through an iterative process, and making presentations about their design process and products. Figure 8.3 shows students, facilitators, and researchers engaged in collaboration, which helps to build the agency of all of these groups.

Throughout this process, all groups share their progress, which provides an opportunity for everyone to see and reflect on how the projects are evolving. We encourage participants to ask questions and to offer and listen to feedback in order to improve their projects. This continues the relationship-building efforts introduced during the launch, deepens communication among participants, and encourages thoughtful feedback. We emphasize that the feedback is meant to benefit the process and that it is up to the designers whether or not to use it. Finally, participants complete their project portfolios, which document their overall experiences in the program.

Figure 8.3 Collaborative relationships to build agency among researchers, facilitators, and participants.

david wells et al.

I2 as an Emergent Learning Environment

All three of the I2 phases blend and overlap, so flexibility in the program is very important. Both the participating youth and adult facilitators collaboratively shape the learning environment; both have opportunities to open up discussion topics and explore tools according to the cohort's needs. This establishes a shared investment in the experience and strengthens the relationship between participants and facilitators. We have experimented with various ways to build trust through language and action. Our mantra is, "Don't follow a plan, follow the teens."

This relationship-driven learning approach is inspired by Vygostky's (1978) theory of social constructivism, which emphasizes the importance of social interaction as the basis for cognitive development. It is also informed by the work of Loris Malguzzi, the founder and director of the preschools in Reggio Emilia, Italy, which have developed an early childhood curriculum that is flexible enough to adapt to the needs of specific groups of students and react to their learning styles and interests throughout the year (Edwards et al., 2011). We used NYSCI's Design Make Play pedagogical approach and created a learning environment that is adaptable to the changing needs of youth. (See this book's Introduction for more details about the Design Make Play principles.) We do this by designing a variety of activities with multiple entry points into the concepts we are introducing. Each week, we reassess and respond directly to the interest level and engagement of participants.

Research Design

We created a research project to deepen our understanding of factors and practices that motivate students to engage in computational making and community-based problem-solving. The research will ultimately examine two implementations of the program, 2019–2020 and 2020–2021. The project tests our hypothesis that, when used together, computational making and social entrepreneurship can encourage youth underrepresented in STEM careers to discover how engineering and design practices and perspectives can be relevant and accessible to them and their communities and can help impact issues that are personally important to them.

Our research during 2019–2020, the first implementation of program, has three goals:

- Study the nature of participants' problem-solving skills and computational making practices throughout the program
- Examine participants' interactions with the materials and people in the program and changes in participants' perspectives about the personal relevance and usefulness of engineering and computational fields
- Design and evaluate research instruments for collecting data on participants' involvement in shaping the program and changing perspectives

To answer our research questions, we developed a research–practice partnership and used existing and newly created research tools to document and study the program.

Research–Practice Partnership

We created a research–practice partnership in which a team of researchers works closely with I2 facilitators and students throughout the program. Our research model is designed to evolve as the program unfolds, in keeping with our mantra to "follow the teens." In this model, the researchers and program facilitators are responsive not only to the needs and interests of the participants but also to each other.

During the 2019–2020 implementation, the team met weekly after each workshop to reflect on the workshop, discuss the challenges and successes in implementing the program and the research, and plan for the following workshop. During these meetings, not only did the design of the program and research components evolve, but the partnership between the researchers and program facilitators also took shape and matured. Each of these groups became more aware of and responsive to the needs of the other group. Below, we describe the research design in more detail.

Research Tools

Based on our integrated model of research and practice, the researchers documented the program week by week to capture elements that might influence student outcomes, in particular the intersection of community problem-solving themes and computational making practices. We used both quantitative and qualitative methods to measure changes in skills and shifts in attitudes, study elements of the program, and examine aspects of students' individual experiences during the program that shaped these outcomes.

Consistent with the program's evolutionary approach, the research team designed and implemented research tools by consistently observing the program, closely interacting with participants, and carefully attending to participants' needs and interests. These tools and techniques were consistently revisited and revised based on feedback from the program team and reflections from the students.

To answer our research questions, we are using multiple data sources:

- *Surveys.* We implemented three surveys (pre-, mid-, and post-participation) to assess students' perceptions about their career paths and their interests in design, engineering, and computational thinking.
- *Videos of workshops.* Using video recordings of the project workshops and written field notes from two embedded researchers, we documented how students identified problems and possible solutions throughout the program. Video recordings were limited to the parts of the program where participants worked on individual or group projects in the museum's Maker Space.
- *Review of curriculum materials with program facilitators.* Weekly discussions with program facilitators provided another source of data to document the evolution of the program.
- *New research instruments.* Soon after we started collecting data and documenting the program, we realized that these data sources could not fully capture the emergent events and learning activities taking shape throughout the program. Therefore, in addition to traditional research instruments, we designed and

david wells et al.

refined new research instruments — a reflection booth and student portfolios, discussed below — to capture the changes that occurred during the program.

Reflection Booth

Based on models that embed assessment into the actual work being done by students (Maltese, 2018) and reflection activities used in our Maker Space, we designed a "reflection booth." There, researchers asked students to reflect on open-ended questions about their challenges, successes, and perceptions of the program in general. We conducted four reflection booth sessions during the year — two during the first semester (exploration phase) and two during the second semester (application phase). In the first semester, the reflection booth was set up in the back of our regular workshop room. Researchers asked students to voluntarily come to the booth at any point during the workshop to talk through their progress. During the second semester, the reflection booths took place online via Zoom since the museum was closed due to the COVID-19 pandemic.

In these guided reflections, we prompted students to think about their process — how they decided on a project idea, what challenges arose during the project, and how they worked through the challenges. In the online reflections, we also prompted students to think about the transition from in-person workshops to virtual workshops and how it was impacting their work. In the final reflection, we also asked students to tell us about their overall experience throughout the whole program.

Students' Project Portfolios

Using Google Sites, we designed a website where the student groups could document their experiences while working on their projects during the second semester. Each student group was asked to design a webpage on the site to document and present their work throughout the project. We provided the groups with a variety of documentation tools, such as Go Pros, video cameras, and an audio recorder, and allowed them to use any format for documenting their work. We encouraged students to not only document their process, but also to think about their collaborative activities, breakthroughs, challenges, and emotions while working on their projects.

Nora's Journey: A Case Study of 2019 Implementation

In the summer of 2019, we kicked off the fifth iteration of the I2 program with a cohort of 17 high school students ages 13–18 (nine males, seven females, and one student who chose not to report their gender). The students were from diverse socio-economic backgrounds, and many resided in NYSCI's surrounding neighborhood. The students participated in this program on a voluntary basis, but 7 out of the 17 students received school credit.

Throughout the program, as participants learned new digital tools and worked on community-centered projects, they simultaneously gained valuable soft skills, which the students themselves described as teamwork, organization, public speaking, and problem-solving skills. Of the students we observed, we noticed that Nora, a 14-year-old girl from Queens, exhibited noticeable shifts in her behavior. (All names of students in this chapter are pseudonyms.) In the first half of the year, for example, Nora rarely participated in group discussions, preferring to interact with her close friend that she met in the program. As the year progressed and students began working on their final projects, Nora assumed a leadership role in her group. She reported increased confidence in public speaking and a deepened understanding of systems thinking and problem-solving processes.

Focusing our preliminary analysis on Nora's experience enabled us to learn more about possible changes throughout the program in students' perceptions of computational tools, their own behavior, and the skills they had learned. Working with Nora also helped us understand students' views about the reasons for these shifts – what outside elements and personal characteristics they believed influenced their growth and evolution throughout the year.

Launch

During the weeklong boot camp in August of 2019, students first built an analog prototype of an educational game using simple materials, tools, and basic circuits with LEDs. They then presented their prototypes, received feedback from their peers, and iterated on their designs. Next, they built a digital version of their game using Ready Maker, a digital tool that enables users without previous coding experience to create games. Nora and her two teammates built a trivia game. We observed that, throughout the group discussions and presentations, Nora remained fairly quiet and often had to be reminded to "speak up."

Exploration

During the first semester of the I2 program, which began at NYSCI in September of 2019, students explored different computation tools, such as Arduino (a platform for building electronics projects), 3D printers, and programming languages. They completed projects, each centered around a different computational tool (see Table 8.1). At the beginning of each workshop, we led students through a brief icebreaker to establish a sense of community and raise energy levels. While working on their projects, we gave students considerable freedom to explore these tools at their own pace and provided them with many materials to experiment with.

Nora worked with a temperature and sound sensor for her Arduino project and used online resources to research how to connect the sensor to her Arduino. Many students, including Nora, described technical challenges related to Arduino. Later in the semester, students explored 3D printers and the TinkerCAD software for creating 3D designs. They designed a GIF of their name using TinkerCAD and then printed out a simple design of their choice on one of the 3D printers. Toward the end of the first semester, a guest speaker introduced students to P5.js, a coding platform for artists and graphic designers. Because she loved art, Nora said that P5.js really appealed to

david wells et al.

Table 8.1 Program modules for the boot camp and first semester

Module	Description
Ready Maker	Prototyped a physical board game; created a digital version using Ready Maker, a digital game design tool
Arduino	Worked on a self-designed project using a microcontroller and a sensor
TinkerCAD	Used TinkerCAD's capability to program circuits; designed and printed a 3D shape
P5.js	Explored how to use code to program interactive, digital art

her and that applying her previous experience with coding made her feel more "confident" in the program. Students also had the opportunity to learn about and use analog tools, including woodworking and soldering tools.

Application

Students started the second semester by making stop-motion animations that told the story of their journey through the first semester. Students went on community walks around Corona, noticing elements of infrastructure and design through the lens of community assets like places of business, worship, recreation, and other gathering sites. Students then spent multiple sessions brainstorming issues in their communities that they were passionate about. From these sessions, four categories for social engagement emerged: school, environmental sustainability, pollution, and cultural disconnection. Each participant prioritized these four categories according to what interested them the most and listed them in their project journals. From these lists we created four groups, one for each theme. Nora chose the cultural disconnection group, as did three of her peers, including Sara, with whom she became good friends over the course of the program.

Three weeks after the students started group work, NYSCI closed due to COVID-19. We discussed how, and if, we could continue the program. We decided to run the program online using Zoom. After some trial and error, we settled into a pattern in which we met on Zoom as a big group and did an icebreaker activity. After this, each group met in a Zoom breakout room. The groups came together again at the end of the session for a project update and sharing session. Transitioning to an online environment was difficult for some students. Nora ably summed up how many of us felt with this observation:

> The literal conversations that we used to have and then the activities we used to do together . . . I feel like I took that for granted and I didn't really think about it that much. And then all of a sudden, we're transitioned into our homes and we're not allowed to go outside at all. And, the whole world is just completely quiet and you're just staring at your screen for, like, over 10 hours.

Some students dropped out due to understandable technical or personal difficulties, but overall, more students than we expected remained committed to the program. All four final project groups had something to present at the end of the program in June. Nora's group continued with their problem statement relating to culture and created a website where people could learn about other cultures and share recipes, art, and music from their own culture.

By looking at multiple sources of data and following Nora's experience firsthand, we came to understand better her evolution through the l2 project. We also came to know her core personal characteristics, which shaped how she moved through new spaces, interacted with other people, and thought about the world around her.

Preliminary Findings About Problem-Solving Practices

Our preliminary analysis focused on students' problem-solving practices – in particular, how students frame approaches for solving problems while working on self-directed projects. We present our findings on two aspects of problem-solving: pursuing self-defined goals and working through challenges.

Defining and Pursuing Goals

Analyzing how students set goals and frame problems is important because these are core practices of problem scoping (Svihla & Reeve, 2016). In addition, goal-setting behavior is closely tied to our larger research questions about student self-efficacy and identity. Students in our program were encouraged to freely explore and devise a project that was personally relevant to them and to select appropriate resources for their projects. To understand the different strategies students use to set their project goals, we studied the transcripts of the reflection booths.

Our analysis of the program's first semester (the skill building and problem-scoping phase) suggests that when students had the time and freedom to develop their own strategy for defining and pursuing project goals, their strategies fell into two main categories: result-oriented and skill-oriented.

Students who used a *result-oriented* strategy devised an idea for a final product before beginning work. For them, skill building was often seen as a means to achieve a final product. Some of these students originated project ideas themselves, while others used online resources to find project ideas. When asked what their initial goal or idea was, these students often responded with, "I wanted to make . . ." or "I wanted to do . . ." For example, Jonah (age 14) explained, "Originally I wanted to do a labyrinth, but I was cocky and I didn't know what I was doing." After trying to replicate a project he found online, he encountered technical challenges with the microcontroller. We observed that other students who, like Jonah, initially had a specific product in mind often had to redefine or simplify their goals as they encountered challenges. Tim (age 16) explained that for his Arduino project he first wanted to use a joystick but soon felt overwhelmed:

> I hooked it up – it was really hard to work with because there were a lot of numbers coming in and I didn't know how to use them . . . So, then I was like, what's the next thing I can use to turn?

Tim held onto the broader project goal, using something to control a motor, but he changed the specific type of sensor after his first idea proved more complex than he expected.

Students who used a *skill-oriented* strategy defined their initial project goal around a skill or concept they wanted to learn. Jesse (age 14) described how he explored different sensors before settling on one project idea: "I was experimenting with other sensors because I didn't really know what an Arduino was . . . So, I guess I wanted to start out a bit simpler, getting the basics done, before transitioning to anything more complex." As this comment illustrates, Jesse recognized that a lack of knowledge prevented him from completing a more complex project. He set a skill-oriented goal in order to later transition to a more complex, result-oriented goal.

Another student, Carlos (age 14), explained that he initially chose an idea for his Arduino project because he wanted to challenge himself:

> I didn't know what most [lines of code] were, so I wanted to learn . . . I was hoping that I would know what these were by the end, and I know that this is a potentiometer and it controls the brightness of the display. I thought this was really cool.

Carlos saw his project as a means to achieve a goal of deepening his understanding about code and microcontrollers. He used code that he found online to program a game with a joystick. Before too long, he was able to get it working in the way he expected. He then spent time adding pieces to his code to make his project more personalized.

Working Through Challenges

Working through an open-ended problem is often very different from the structured problem-solving approaches students experience at school. Much less is known about how young people make sense of complex, self-defined problems. Our analysis found that although students frequently described challenges they faced throughout the workshops, they had different interpretations of what constituted a challenge. For example, some students described challenges with learning how to use a new tool or software, whereas others identified challenges with social and emotional experiences. Below, we briefly discuss the main types of challenges that emerged from our data.

Some students faced challenges with initiating an idea and translating it into a project. As Alex (age 15) noted, "We always have an expectation in our heads, so what's always challenging is how to put what we think on the paper. It's very hard for me to express what I think into the project." Before Alex started work on his design in TinkerCAD, he said he had a very clear idea of what he wanted the final result to look like but knowing how to realize that vision was "challenging." However, he also mentioned that his favorite moments of the program occurred right after facilitators introduced a new tool, when he was able to "think about all the possibilities." This suggests that although open-ended projects create challenges for students in generating and realizing their own ideas, these challenges can motivate students to explore further.

Students frequently mentioned difficulties with troubleshooting – understanding a new system and "getting it to work." During one reflection, Tim described the challenge of identifying an underlying problem when working with an unfamiliar system because mistakes may not be immediately recognizable. As Tim explained, "You have to trace back your steps – I think learning how to do that is really important, but also really hard to teach because you've just gotta make those mistakes and figure out how to do it afterwards." Tim's comment supports our claim that a loosely structured program grants students opportunities to practice important problem-solving skills. Many other students described similar processes in which they recognized their frustration while troubleshooting problems but also understood the importance of going through this process.

Some students described social and emotional challenges, such as staying motivated, navigating frustrations, and working on a team. Anju (age 13), for example, emphasized the importance of keeping calm as she worked through a problem with her Arduino: "Instead of being angry and accidentally breaking the sensor, I tried to stay calm, and at that point I was thinking, 'I have to keep on trying. I have to make this work – I can't just give up.'" When asked what motivated her to persevere, she said that knowing that her classmate was able to get the same sensor to work helped her persist. She also mentioned that it was her second attempt at working with a sensor, so there wasn't enough time to start over. Those two factors, peer support and limited time, "fueled [her] not to give up."

Other students, rather than focusing on a challenging tool or skill, described challenges of confronting preconceived ways of thinking or doing things. Some of the students noted that the lack of scaffolding and initial instruction to help them work with new tools or systems was different from what they experienced at school. Anju made the observation that learning how to work with a 3D printer "showed how complex something can be when you don't have a template. In school, usually there's a model for you . . . Here, it's in your hands; you have to be really independent." Anju described this as a positive challenge and emphasized that he felt powerful taking control of a computational system. Carlos found it challenging to create art using technology. He explained that, although he normally has control over his hands, when he uses technology he has to "tell it what to do." Although he concluded that it's easier to make art with one's hands, he said that thinking about making art with technology was a "cool challenge."

Strategies for Working Through Challenges

We found that students approached challenges in different ways and used a variety of strategies to work through challenges. These strategies fall into three main categories:

- *Researching a problem.* Many students used online resources to research possible ways to fix technical problems that arose; these resources included YouTube videos, help forums, and curated online resources sheets provided by facilitators.
- *Troubleshooting.* Many students demonstrated troubleshooting practices, particularly to address challenges that arose with the Arduino system. Students often described this process as "trial and error."

david wells et al.

- *Asking for help.* Many students worked through challenging problems by asking their peers or facilitators for assistance. Almost all students worked one-on-one with a facilitator at some point during these projects. Before reaching out to facilitators, many students said they first turned to their peers.

Conclusion

Self-efficacy – how we perceive ourselves and our capabilities – can support us through the toughest times and leave us unbearably vulnerable at other times (Bandura, 1997). When students expose their vulnerability in an environment of guidance and support, this can help them get over potential bumps and feel more positive and proficient. Having confidence in one's ability to approach new experiences and feel successful as a learner helps to fuel the desire to go on learning (Dewey, 1938).

In the I2 program, we strive to provide many opportunities for students to develop agency by exploring STEM competencies and career pathways. We welcome diverse thinking and hope to inspire further complexity and create new tools for thought. The entry point for students to engage in I2 must be designed, repurposed, and adjusted as the program increases in complexity. We have also found that providing choice is an effective engagement strategy. Providing choice gives participants an opportunity to actively mold their experiences and understand that their voice matters. This helps to develop a sense of shared responsibility among everyone in the program and creates cohesion within each cohort. Embedding these three aspects – self-efficacy, invitation, and choice – into the program creates an inclusive environment for participants of all skill levels and expressive styles that defines the Innovation Institute experience.

Acknowledgments

This material is based upon work supported by the National Science Foundation under Grant No. 1759261. The authors would like to thank additional Maker Space staff and researchers who were involved in this project, including Lauren Vargas and Betty Wallingford. We would also like to thank all of the participants in the Innovation Institute program for their engagement and contribution to the development of this program. We couldn't have done it without you!

References

Baily, T. C. (2015). *Organizational support, satisfaction, and STEM research career plans in pipeline interventions: A strengths-based approach among underrepresented students* [Doctoral dissertation, University of Michigan]. https://deepblue.lib.umich.edu/bitstream/handle/2027.42/111494/tashab_1.pdf;sequence=1

Bandura, A. (1997). *Self-efficacy: The exercise of control*. W. H. Freeman.

Benda, K., Bruckman, A., & Guzdial, M. (2012). When life and learning do not fit: Challenges of workload and communication in introductory computer science online. *ACM Transactions on Computing Education, 12*(4), 15.

Boy, G. A. (2013). From STEM to STEAM: Toward a human-centred education, creativity & learning thinking. In P. A. Palanque, F. Détienne, & A. Tricot (Chairs), *ECCE '13: Proceedings of the 31st European Conference on Cognitive Ergonomics* (pp. 1–7). Association for Computing Machinery (ACM). https://doi.org/10.1145/2501907.2501934

Buechley, L., Eisenberg, M., Catchen, J., & Crockett, A. (2008). The LilyPad Arduino: Using computational textiles to investigate engagement, aesthetics, and diversity in computer science education. In M. Czerwinski & A. Lund (Chairs), *Proceedings of the ACM CHI Conference on Human Factors in Computing Systems* (pp. 423–432). ACM. https://doi.org/10.1145/1357054.1357123

Buechley, L., Peppler, K., Eisenberg, M., & Yasmin, K. (2013). Textile messages: Dispatches from the world of e-textiles and education. *New Literacies and Digital Epistemologies: Book 62.* Peter Lang Publishing Group.

College Board. (2015). *AP computer science principles, 2016-2017: Curriculum framework.* College Board. https://apcentral.collegeboard.org/courses/ap-computer-science-principles

Dewey, J. (1938). *Experience and education.* Kappa Delta Pi.

Eccles, J. (2009). Who am I and what am I going to do with my life? Personal and collective identities as motivators of action. *Educational Psychologist, 44*(2), 78–89.

Eccles, J. S., & Wang, M. T. (2015). What motivates females and males to pursue careers in mathematics and science? *International Journal of Behavioral Development, 40*(2), 100–106.

Edwards, C., Gandini, L., & Forman, G. (Eds.). (2011). *The hundred languages of children: The Reggio Emilia experience in transformation* (3rd ed.). Praeger.

Ellis, J. M. (2015). College readiness beliefs and behaviors of adolescents in a precollege access program: An extension of the theory of planned behavior [Doctoral dissertation, University of Michigan]. https://deepblue.lib.umich.edu/bitstream/handle/2027.42/116736/jmelli_1.pdf?sequence=1&isAllowed=y

Kafai, Y. B. (2006). Playing and making games for learning instructionist and constructionist perspectives for game studies. *Games and Culture, 1*(1), 36–40.

Kilgore, D., Atman, C. J., Yasuhara, K., Barker, T. J., & Morozov, A. (2007). Considering context: A study of first year engineering students. *Journal of Engineering Education, 96*(4), 321–334.

Lauermann, F., Chow, A., & Eccles, J. S. (2015). Differential effects of adolescents' expectancy and value beliefs about math and English on math/science-related and human services- related career plans. *International Journal of Gender, Science and Technology, 7*(2), 205–228.

Maltese, A. (2018). MakEval: Tools to evaluate maker programs with youth. www.adammaltese.com/content/makeval/

Martin, R. L., & Osberg, S. (2007). Social entrepreneurship: The case for definition. *Stanford Social Innovation Review, 5*(2), 28–39.

National Research Council. (2010). *Report of a workshop on the scope and nature of computational thinking.* The National Academies Press.

Papert, S. (1993). *The children's machine: Rethinking school in the age of the computer.* Basic Books.

Peppler, K. (2013). STEAM-powered computing education: Using e-textiles to inte-grate the arts and STEM. *IEEE Computer, 46*(9), 38–43.

Peppler, K., & Glosson, D. (2013). Stitching circuits: Learning about circuitry through e-textile materials. *Journal of Science Education and Technology, 22*(5), 751–763.

Qiu, K., Buechley, L., Baafi, E., & Dubow, W. (2013). A curriculum for teaching com-puter science through computational textiles. In N. Sawhney, E. Reardon, & J. P. Hourcade (Chairs), *Proceedings of the 12th International Conference on Interaction Design and Children* (pp. 20–27). ACM.

Rode, J. A., Weibert, A., Marshall, A., Aal, K., von Rekowski, T., el Mimoni, H., & Booker, J. (2015). From computational thinking to computational making. In K. Mase, M. Langheinrich, & D. Gatica-Perez (Chairs), *Proceedings of the 2015 ACM International Joint Conference on Pervasive and Ubiquitous Computing* (pp. 239–250). ACM.

Svihla, V., & Reeve, R. (2016). Facilitating problem framing in project-based learning. *Interdisciplinary Journal of Problem-Based Learning, 10*(2). https://doi.org/10.7771/1541-5015.1603

Vossoughi, S., & Bevan, B. (2014). *Making and tinkering: A review of the litera-ture.* Commissioned paper for the Committee on Successful Out-of-School STEM Learning: A consensus study. National Research Council. https://sites.nationalacademies.org/cs/groups/dbassesite/documents/webpage/dbasse_089888.pdf

Vossoughi, S., Hooper, P., & Escudé, M. (2016). Making through the lens of cul-ture and power: Toward transformative visions for educational equity. *Harvard Educational Review 86*(2), 206–232.

Vygotsky, L. (1978). *Mind in society.* Harvard University Press.

Wang, M. T., Eccles, J. S., & Kenny, S. (2013). Not lack of ability but more choice: Individual and gender differences in choice of careers in science, technology, engineering, and mathematics. *Psychological Science, 24*(5), 770–775.

Wardrip, P., & Brahms, L. (2015). Learning practices of making: Developing a frame-work for design. In M. Umaschi Bers & G. Revelle (Chairs), *IDC '15: Proceedings of the 14th International Conference on Interaction Design and Children* (pp. 375–378). ACM. https://doi.org/10.1145/2771839.2771920

Weintrop, D., Beheshti, E., Horn, M., Orton, K., Jona, K., Trouille, L., & Wilensky, U. (2016). Defining computational thinking for mathematics and science classrooms. *Journal of Science Education and Technology, 25*(1), 127–147.

Wigfield, A., & Cambria, J. (2010). Students' achievement values, goal orientations, and interest: Definitions, development, and relations to achievement outcomes. *Developmental Review, 30*(1), 1–35.

Wigfield, A., & Eccles, J. S. (2000). Expectancy-value theory of achievement motiv-ation. *Contemporary Educational Psychology, 25*(1), 68–81.

Wing, J. M. (2008). Computational thinking and thinking about computing. *Philosophical Transactions of the Royal Society of London A: Mathematical, Physical and Engineering Sciences, 366*, 3717–3725.

Part III
Playing and Learning Across Settings

See, Touch, and Feel Math

Digital Design for English Language Learners

Dorothy Bennett, Tara Chudoba,
Xiomara Flowers, and Heidi Slouffman

Overall, I am surprised by how [my student] has shown me his higher level of engagement, which made me realize that he in fact is not really shy but just cautious. His seemingly distracted demeanor in my class was not because of his lack of engagement but rather because of his lack of English proficiency. He needed to constantly validate his work with his seatmates. This activity made me realize that ELL students like [this student] needed more visually engaging activities that can help neutralize the challenges and problems of English learners in regular classroom settings.

A teacher participant in Digital Design for English Language Learners (ELLs) at the New York Hall of Science (NYSCI)

Digital Design for English Language Learners was an out-of-school pilot program designed to address the mathematics and language literacy needs of the fastest-growing population of students in the United States. Our specific focus was our own community of middle schoolers living in Queens, New York – the linguistic capital of the world, where more than 138 languages are spoken (Solnit & Jelly-Schapiro, 2016). In partnership with public school teachers from a neighborhood middle school, educators and researchers from NYSCI developed and tested design-based instructional activities and strategies that aimed to support "math talk" and hands-on exploration of hard-to-teach mathematics concepts. We also provided professional development opportunities for local middle school math and ELL teachers.

The activities centered on the use of two of NYSCI's Noticing Tools: Choreo Graph and Fraction Mash. These apps invite learners to see, touch, and use mathematics to design personalized digital artifacts. Reflecting this multimodal approach, we created a program that invited ELLs to create digital design projects that connected to their lives and cultures and to engage in activities that involve the whole body, referred to as "embodied learning" (Lee, 2015). Along these lines, learners physically acted out, discussed, experimented with, and represented the mathematical ideas behind their designs.

Based on lessons learned from this work, this chapter discusses the new forms of engagement that can emerge when formal educators who are experts in working with ELL students are invited into informal learning environments. It also describes how combining human resources, creative digital tools, and materials in innovative ways can promote a sense of agency and inclusion in mathematics learning for populations with needs and strengths that often go overlooked in formal education settings. This work further demonstrates how science museums can serve as a "third space" in which teachers, students, and their families learn alongside each other about science, technology, engineering, and mathematics (STEM) and how a low-stakes but engaging environment can provide rich, expressive opportunities for communication and math exploration. Finally, the chapter sheds light on how codesigning and testing activities with educators in an informal learning environment can have a profound impact not only on the participating students, but on their teachers, as they see how more open-ended, informal, design-based approaches provide a means for identifying and building on the assets that ELLs bring to the learning experience.

Why Digital Design?

There is much debate about the best approaches for addressing the needs of students whose first language is not English and the criteria for identifying which students are ELLs. Across the spectrum of approaches, however, it is widely acknowledged that in STEM learning, ELLs are often confronted with the dual challenges of grasping the nuances of a new language while learning core academic content with a specialized vocabulary. Leading research suggests that some of the best instruction provides diverse opportunities for speaking, listening, reading, and writing while encouraging learners to take risks, construct meaning, and seek reinterpretations of knowledge in different contexts (Garcia & Gonzalez, 1995). Models, visual representations, pictures, manipulatives, and problems that connect to students' lives and cultures are bridges that help students understand core content ideas while deepening their language skills (Moschkovich, 2012; Quinn et al., 2012). They also benefit from thematically integrated projects that promote higher-order thinking, cooperative learning, and high-quality exchanges between teachers and students (Hindley, 2003; Moschkovich, 2012).

We speculated that two of NYSCI's Noticing Tools, which emphasize hard-to-teach mathematics concepts, visual design projects, and storytelling, would be well suited to addressing these needs and bringing together many of these strategies for teachers and ELL students. Inspired by dance and choreography, Choreo Graph is an iPad app that enables users to create animated characters from photos and employ mathematical concepts and practices to help choreograph their chosen subjects' motions. Fraction Mash is an iPad app that allows users to create playful photo mashups, such as swapping parts of a friend's face with a cat's, while reasoning about foundational concepts of fractions, such as how fractional parts relate to the larger whole of the image being created. Both tools allow children to readily incorporate their interests by taking photos of the people, places, and things they care about to create new digital artifacts and interactive stories. Both also encourage children to use measurement tools and mathematical functions to help realize and improve their designs.

Drawing on prior research, we set out to conduct two week-long pilot workshops outside of school time to create a set of supplemental resources that could support ELLs' mathematics and language literacy development through design-based approaches. We aimed to generate and test strategies and activities that would achieve these goals:

- Support ELLs in using mathematics as a tool for problem-solving
- Broaden opportunities for students to engage in "math talk" and promote language literacy
- Improve teachers' abilities to assess ELLs' mathematical thinking through observation and recording
- Ultimately develop instructional strategies for using Noticing Tools with ELLs to foster mathematics and language literacy

We faced a challenge, however. While prior work told us that our digital tools were an expressive, engaging, and exciting way for students to develop mathematical fluency, teachers did not always see how they could leverage the tools and integrate them into their already compressed curriculum. It became obvious to us that teachers needed more support with figuring out how to allow adequate time for exploration and play and integrating the tools into their teaching practices and math instruction for ELLs.

To address these needs, we collaborated with two seasoned ELL teachers to identify the potential of the tools for ELLs' learning and to co-create a set of complimentary non-digital activities for deepening mathematics discourse and exploration. We invited them to join us in running the workshops for local students in our museum – a low-stakes environment where play and exploration are valued and where the usual concerns and restrictions of formal classrooms can be suspended for a short time. We were committed to having teachers facilitate activities alongside the museum staff. Rather than requiring a specified curriculum, we encouraged our teacher partners to experiment with inventive ways to meet the core content learning goals of the middle grades' curriculum. We also wanted teachers to have the chance to help children explore mathematical ideas more deeply through design approaches and to witness the serendipitous discoveries children often make and share when they are left to play and explore more freely.

This was a bold agenda. We would need to find teachers who were willing to go along with this iterative design process.

A Formal/Informal Partnership

An important aspect of this work was the interdisciplinary team we forged of people from formal and informal learning settings. The team included three NYSCI researchers and educators familiar with design-based learning approaches. Their role was to guide the overall development of the workshops, introduce teachers to the tools and accompanying resources NYSCI had already developed through prior efforts, and document the work of students and teachers during the workshops. Other team members included lead ELL teachers who codesigned and facilitated the workshops, teacher-observers from our local neighborhood middle schools, ELL students, and multilingual youth Explainers, who facilitate exhibits on the museum floor. These collaborators, described below, brought unique skill sets and expertise and provided significant input and insights that shaped the design of the program.

Lead ELL Teachers

Two ELL teachers were recruited through NYSCI's existing teacher networks to serve as lead teachers on the project. These lead teachers collaborated with NYSCI staff to organize the pilot workshops, codesign activities, and plan each week's schedule. They also co-led the workshops with support from our Explainers.

The lead ELL teachers were chosen based on their different approaches to ELL instruction and their openness to experimentation. The first is a bilingual educator with 28 years of experience who began the first bilingual education program for the Uniondale Union Free School District on Long Island. In this district, 50% of the students speak a language other than English. This instructor came to the United States from Panama when she was 12 years old, and her passion for bilingual education comes from her experience as an ELL in middle school on Long Island. Previously, she taught in Corona, Queens, where NYSCI is located. As a result, she has a deep sense of empathy for ELL students and related personally to many of the students participating in the program.

The second lead is a TESOL (Teaching English to Speakers of Other Languages) teacher for fifth graders in an Astoria, Queens, public elementary school. Astoria, one of the most diverse neighborhoods in the country, closely reflects the community that our museum serves. This instructor has been teaching English and mathematics for over 13 years and developed the first English as a Second Language (ESL) program for her prior school in Brooklyn. She was accustomed to using a broad set of strategies in her classroom for regularly communicating across different languages with her students.

While both teachers used instructional technology, they were not accustomed to using open-ended digital tools like those developed at NYSCI. Before the first workshop, the lead instructors and NYSCI staff participated in seven design meetings

dorothy bennett et al.

to learn about the tools, refine our existing supplemental activities and create new ones, and plan how the workshop would unfold. Before the second workshop, they met three times to revise activities and workshop schedules based on the previous session's results.

Teacher Observers

To better develop activities that would build bridges between informal approaches to exploring mathematics and language literacy and the more formal mathematics and English language arts instructional practices used in school, we invited teachers from our partner middle school in Corona, Queens, to observe their ELL students in our workshops. We asked them to watch for and reflect on the mathematical reasoning and language literacy demonstrated by their students as well as anything that surprised them. Five teachers were actively involved in recruiting students, and four of them came to observe the workshop. These were fully certified teachers of a range of subject matters (mathematics, biology, ESL, TESOL, special education). Leading up to the pilot workshop, the project team hosted professional development sessions to familiarize the teacher-observers with the Noticing Tools and train them on an observation protocol created with the help of expert advisors. The protocol was designed to guide the teachers in looking for evidence of student learning and engagement, taking notes on what was happening, writing about what they saw, and describing instances of students expressing mathematical thinking through non-verbal behaviors, such as counting on fingers or gesturing to illustrate concepts like greater than and less than.

Middle School ELL Students

Across both workshop sessions, 19 students from diverse backgrounds were recruited to attend a "fun math and design workshop" by teachers in our partner middle school. Students spanned a range of experiences and English proficiency levels. Of the 19, 5 had come to the United States within the previous six months. The remaining 14 students had intermittent residency in the United States; many of them were born in the US but travelled back to their family's home country, sometimes for two to three years at a time. The majority were Spanish-speaking students with heritages from Mexico, Ecuador, and the Dominican Republic. Two of the students spoke Hindi and Punjab, and one of the US-born students was on the autism spectrum.

Explainers

NYSCI regularly employs high school and college students drawn from all boroughs of New York City to serve as Explainers in our museum. Explainers work as floor facilitators for our exhibits and often assist with educational programming. Two multilingual Explainers were recruited to set a playful tone for each workshop and to create an inviting and accessible space for ELLs. The Explainers assisted with translations and encouraged and supported students by leading daily icebreakers that set the tone for fun and learning. Their involvement helped us to get to know the personalities and general English proficiency levels of the students and to obtain cogent feedback on our programmatic work.

Designing Multimodal Approaches to Foster Agency and Inclusion

Our pilot program sought to develop and test a set of activities in which students, regardless of their English language proficiency, could express their mathematical ideas through different modes such as conversing, writing, and debating in their preferred language, as well as drawing, designing, animating, and physically acting out concepts and ideas. Toward this end, the team described above worked collaboratively to codesign an intensive workshop where middle school students with varied math and English language facility could engage in physical and digital design activities. These activities, which centered on NYSCI's Math Noticing Tools, would give students opportunities to explore and apply a range of foundational mathematical concepts and practices, such as fractional parts of a whole, angles of rotation, coordinate geometry, symmetry, and translations. Our expert teacher instructors worked to foster mathematical discourse through hands-on activities and more well-established strategies for supporting ELLs' language development.

The general daily format included a mix of introductory full-body activities; digital app-based design projects targeting a certain concept or theme, like fractions of a student's personality; design journals in which students could sketch ideas or reflect on their thinking in their preferred language; and small and large group discussions where they presented and debated the math behind their designs. Student-defined digital design projects enabled participants to more deeply explore the math concepts and practices mentioned above to tell their own stories – for example, by creating animated amusement park rides involving all their friends. At the end of the week, students invited their families to a celebration of their work in which they proudly presented the projects and explained the mathematics used to create them.

A series of activities on symmetry illustrates this multimodal approach, which was intended to ensure that students of all abilities and strengths would experience success. With help from instructors, students started modeling different types of angles with their arms. Then students acted out and videorecorded their own angle dance moves. They used jump ropes to find their lines of symmetry on their own bodies while striking a pose so their partner could dissect how the move was symmetrical. This was followed by more individual design work in the Choreo Graph app; students created their own characters and animated a symmetrical dance, and then presented their animation to fellow workshop participants and debated with the group whether or not their animated dance was truly symmetrical. At the week's culmination, students created animation projects that put together the core ideas (both mathematical and linguistic) they had learned during the workshop activities and focused on themes of personal interest.

The digital design workshops were held twice at NYSCI, the first during the New York City Department of Education's spring break and the second during the summer break in 2017. Throughout the week the students made use of the museum, visiting exhibits

dorothy bennett et al.

for inspiration, taking photos for their projects, going on scavenger hunts for angles on NYSCI's Science Playground, and exploring symmetry in our gravity mirror exhibit and in the Noticing Tools. Pilot data from the first workshop were used to revise the activities for the second workshop. Throughout the development process, the pilot program's advisors – well-respected experts in math instruction for diverse communities and ELLs – were enlisted to review our tools, advise us on potential activities and embedded assessments, and guide our work with the teacher-observers. Based on the two workshops, the project team developed a set of supplemental resources to help informal and formal educators use the Noticing Tools and design-based activities to support ELLs in mathematics and literacy.

ELLs' Experiences With Math and Language Literacy Through Digital Design

To better understand the impact of our co-created multimodal design activities on student engagement with mathematics and language literacy, all sessions and discussions were observed by a researcher, and group and paired discussions were videotaped for later review. Students kept reflection journals, participated in pre- and post-embedded assessment activities on fractional understandings, and participated in post-workshop interviews about their experiences in the workshops. We also interviewed teachers and parents about their perceptions of the program's impact on themselves and their children. Triangulating the data, we analyzed the differential impacts of the workshops on all participants.

Below we share three cases that exemplify how the program addressed the needs of students with different levels of math and English literacy and promoted agency and inclusion for students with a broad range of strengths and needs.

Mariana: Finding Confidence and Voice in a Supportive Environment

Sixth-grader Mariana, who speaks primarily in Spanish, came to the Digital Design Workshop because her mother wanted something for her to do during spring break. (All student names in this chapter are pseudonyms.) Mariana had moved to the United States from Mexico three months prior. At the beginning of the workshop, she was very hesitant to speak to others. The instructors and Explainers did not insist that she use English but instead used Spanish to bring out her project ideas and encourage her mathematical thinking. The bilingual instructor and Explainers translated to help her follow what was happening in the workshop and to support her in sharing her ideas with English speakers.

Before the workshop, she had liked math but thought the way it was taught in school, with repetitive exercise, was boring. Mariana also confessed that in school she struggled to understand the lessons, even though she had learned much of the content in Mexico, because students and teachers could not take time to explain things to her a second time or translate for her. In the NYSCI pilot workshop, her grasp of mathematical concepts and skills became quickly evident in the complex designs she created. Her designs gave instructors and peers a window into her thinking and planning process.

In an activity focused on geometric translations, Mariana used the coordinate grid and translation visualization tools in the Choreo Graph app to create a dual-figure symmetrical animation with translation paths (see Figure 9.1) that formed two trees. In her journal she wrote down the corresponding coordinates of the first figure she created to precisely recreate the other tree on the other side. This design was more complicated and required more precision and spatial reasoning than the original activity prompt called for. This and other designs she created garnered the attention of other students, who were eager to find out how she had done them. The visual medium of Choreo Graph gave Mariana the opportunity to share her knowledge of coordinates, translation, and symmetry with members of the group who did not speak Spanish, even though she could not yet express these ideas in English.

Over the course of the workshop, Mariana primarily worked with two other girls who spoke some English and Spanish. She quickly gained authority for her mathematics knowledge and her ability to design with the app and was frequently consulted for help. Mariana continued to take on challenging design projects. She gained confidence in presenting in Spanish to the larger group the mathematics she used while others in her group translated. For her final project, Mariana and the other girls in her trio created a three-part amusement park animated story. Mariana's portion of

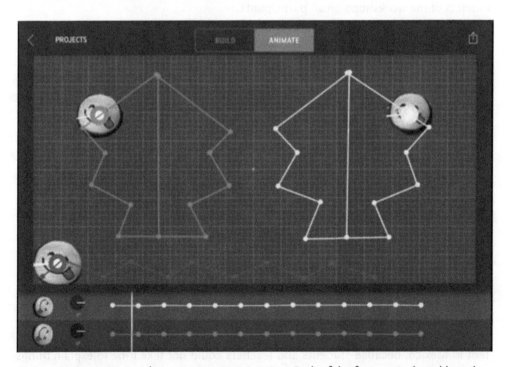

Figure 9.1 Mariana's translation symmetry animation. Each of the faces was plotted based on coordinates she had written down to be precisely symmetrical.

dorothy bennett et al.

the project was complex: she created a rotating Ferris wheel and animated the figures so that they each boarded the Ferris wheel one-by-one and then rode on it (see Figure 9.2). She struggled with but eventually succeeded in making the Ferris wheel rotate in place, separating the rotation of the wheel from the translation of the figures boarding it. She used her coordinates and angle of rotation tools to synchronize each character boarding the Ferris wheel.

After the workshop, Mariana declared that she found math much more enjoyable and would prefer to attend the workshop over her everyday school:

> Well, I think that math is very good. Before, I thought math was very boring because they taught us in a normal way. But here they teach us in a very different way. I say it's very fun, and if they asked me to choose to go to school there or come all the days to school here, I would choose the school here.

In the safe space of the workshop, Mariana was able to express prior and newly acquired math knowledge with instructors who responded with patience and interest. By midweek, it was clear that Mariana had also gained confidence in her ability to communicate in English and often approached the workshop Explainers to practice speaking it. She said that the type of instruction and the use of the Noticing Tools allowed her to relate math to her everyday life and better understand it in a shorter period of time than she could do at school.

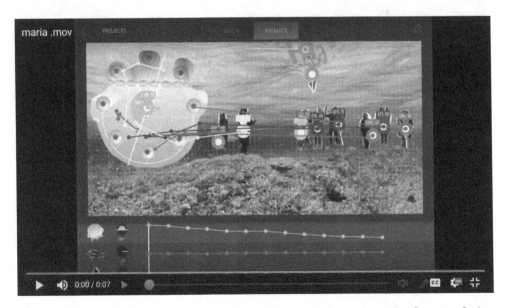

Figure 9.2 Mariana's Ferris wheel project. The top line graph shows the angle of rotation for her Ferris wheel decreasing at each time point, making it slowly rotate as each character boards the wheel.

Alejandro: Learning Through Design and Physical Play with Family and Peers

Alejandro was born in the United States but moved back to Mexico for five years; he then travelled back by himself to join his family in New York. He attended the last portion of fifth grade in New York the year the workshop took place and was starting his first full school year as a sixth grader in the fall. He came to the workshop with his big brother, who was a rising ninth grader; although the brother had not intended to stay at the workshop, he decided to remain for the week to do his own projects. At the start of the workshop, Alejandro seemed overwhelmed. He had difficulty settling into the workshop environment and the pre-assessment task and left his pre-assessment mostly blank. He rarely spoke, preferring instead to nod or shake his head if the question allowed it. The hands-on activities and open-ended project work seemed more effective in drawing him into the workshop than direct questioning or pushes for conversation did. During the open-ended work, Alejandro opened up the most; he talked about his projects and personal interests and occasionally smiled and laughed.

Using the Fraction Mash app, Alejandro created interesting photo mashups, rotating both images upside down and merging them (Figure 9.3) or placing his face on a dollar bill. As part of a group activity, workshop members were asked to find

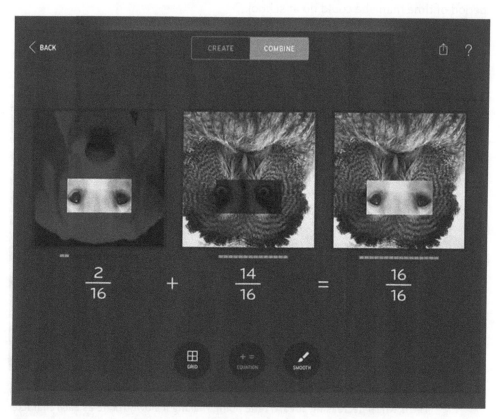

Figure 9.3 A dog and an owl merged together using the Fraction Mash app.

dorothy bennett et al.

a fractional part in their fraction designs and determine where it might go on a big number line on the floor that went from 0 to ½ to 1. While at first Alejandro struggled with this, the instructor playfully walked him through the process of locating his particular fractional mashup on the number line and with each guess asked, "What about here, larger or smaller than one half?" This encouraged him to try out new mashups and proudly find their values along the line on his own. Soon thereafter, another Spanish-speaking student was paired up with Alejandro to figure out how to find half of an odd number (e.g., what is one-half of 17?) after they had just confronted this task in their project work. This allowed Alejandro to discuss a difficult idea with a male peer.

As the week progressed, our bilingual teacher made headway with Alejandro by explaining to his older brother that "just like you had to practice soccer, your brother will acquire his new targeted language." She reminded the brother that he would be a great support in helping Alejandro practice and "apply his first language of Spanish to bridge to English." With his brother's encouragement, it was not long before Alejandro got involved in more complex design projects. Exploring dances to convey a mood in Choreo Graph, Alejandro intensely focused on creating a multiple-jointed animation (Figure 9.4). He put vertices in each joint of the limbs of his character to create realistic movements, and he attempted to match the angles at each of the ten joints on

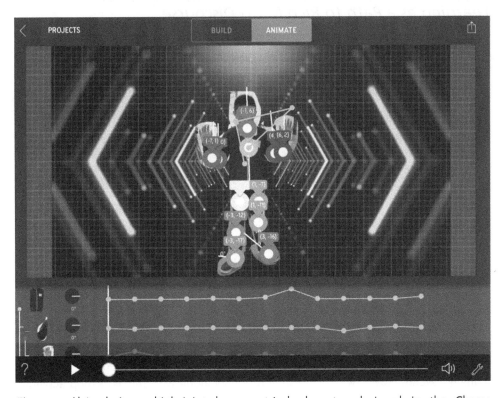

Figure 9.4 Alejandro's multiple-jointed symmetrical character designed in the Choreo Graph app.

his character's body. With his brother, he also created some of the most challenging "symmetry or no symmetry" photos for the class to decipher, creating parallel poses that encouraged the entire class to debate their symmetry.

About midweek, Alejandro attempted fuller presentations of his work. When facilitators showed genuine interest and curiosity, he responded with quiet explanations and even made full presentations in English in front of the group and during the family celebration. For his final presentation, he wrote out a narrative in English about his complex Mario game recreation, explaining how he used angles to guide the movement of Mario and other characters in his animation. In a post-interview, Alejandro said that his proudest moments were when he got in front of the workshop members to explain his work and how he felt much more confident in presenting his work. After the family celebration, his mother came forward to thank the workshop staff and its facilitators:

> I want to thank all the teachers that have taught new things to my boys and had patience, more than anything, with [Alejandro], because he arrived six months ago, and it's really difficult for him . . . the language more than anything. Thank you very, very, very much for the support . . . and the confidence that you all gave to him.

Rayyis and Nikhil: Cultural Connections and Paired Learning as a Path to Rich Math Discourse

Rayyis and Nikhil were unique in these workshops, as they were the only two ELLs who did not speak Spanish as their first language. While Nikhil was from Nepal and Rayyis was from Bangladesh, both had been in the United States for only five months and were encouraged to help each other navigate a workshop where no one else spoke their languages. Rayyis spoke Urdu, with slightly more command of English than Nikhil. Nikhil spoke Bengali, a dialect of Punjabi that could be understood by Rayyis even if it differed slightly. Despite still learning English, Nikhil was more verbal during the design process than Rayyis and pushed his partner to create more ambitious work. The two paired up on a single iPad that Rayyis primarily controlled, even though each had his own iPad to work on.

Our instructors were initially challenged by these students. Our TESOL-trained instructor decided to help them by encouraging them to physically act out concepts, such as forming obtuse angles with their arms. Rayyis and Nikhal embraced the physicality of the instruction, referring back to their enactments later in the workshop. They also relied heavily on the charts and written posters on the walls to find English math terms for the concepts they explored in the workshop so they could communicate to others what they had done.

From the beginning of the workshop, Rayyis engaged in group discussions and debates bilingually. He became a leader for the pair, presenting the work that he and Nikhil had put together and translating for Nikhil when they presented their work individually. Overall, Rayyis was better able to follow along with the workshop activities and often told Nikhil what to do during the workshop. Nikhil followed Rayyis's lead but did not do so unquestioningly. When Rayyis would get frustrated with something

dorothy bennett et al.

they were trying to tackle in a project, Nikhil would persevere and question Rayyis's thinking or correct his calculations when they were off. The two engaged in spirited debates about interpretations of the mathematics they were using. This was also reflected in their journals and ultimately resulted in greater understanding for both of them.

Nikhil had many ideas of his own throughout the workshop, but due to his limited experience with English he could not directly communicate them to the instructors or fellow workshop participants. Instead, he brought his ideas into the products that he and Rayyis created together and would make clear during presentations which ideas he had contributed. Through bilingual conversations during their collaborative process, the two students invited facilitators and instructors into the discussion. Facilitators guided them in working together, as Nikhil made his ideas clear for what should happen and Rayyis executed them or pushed back in favor of his own ideas. At the end of the process, Nikhil often chimed in with the English he knew to answer the instructor's questions about their math reasoning, relying on word walls and charts around the room. While Nikhil was unable to make it to the final day's family celebration, Rayyis presented some of his collaborative work with Nikhil and gave a detailed account of how he constructed an animated soccer game for his final project featuring him and his family. After this presentation in English, Rayyis's father expressed pride in his son's progress: "Do you realize he is only in the country six months? I can't believe that he was able to be up in the front and do this."

What Agency and Inclusion Look Like for ELLs and How to Get There

As these student cases reveal, agency can look different for ELLs than it does for native English speakers. While an on-task native English speaker may listen and respond to questions at the time a teacher asks them, students who are still learning the language may need to translate for each other if no other translation is available. Students may also be unusually quiet as they are processing what they hear; they may have to translate the English in their heads before answering and asking questions. In the first pilot session, one student mentioned in an interview that in school she often did not feel supported because her teachers and classmates acted as if she already understood English.

Throughout the workshops, all students progressed in their abilities to talk about their work and the math they were using, both in academic and non-academic terms (e.g., bigger, smaller, and the same). The students who learned math concepts in Spanish were motivated to learn English terms through more conventional methods like word walls that our instructors had placed around the room. Many said they felt they were in a supportive environment where they could succeed because they could understand what was happening. Students also worked with each other bilingually and in their first languages to help each other figure out problems and push each other's mathematical thinking and design work.

The apps were a useful tool for working through misconceptions. Teachers encouraged students to experiment with and test their working assumptions within the app. For example, if students thought that the larger the denominator in a fraction, the bigger the fractional part, they were invited to directly manipulate the denominators of their mashups in the app and watch as the fractional parts grew or shrunk. While they explored, facilitators asked open-ended questions about what students noticed, what relationships they were seeing, and how they could test their theories further in the app. When the class later engaged in discussions about their theories, instructors repeatedly asked the students to share the reasoning behind their claims or use evidence they had found in the app to illustrate their points. Relatively quickly, students began offering their reasoning without prompting. These same students often continued to share their mathematical discoveries with others, using gestures, academic vocabulary, and nonacademic descriptions to discuss ideas related to their designs. By encouraging them to consult with one another or show off what they designed with the Noticing Tools, facilitators fostered students' agency over their own learning.

Perhaps the most successful strategy was the use of kinesthetic modeling and learning. The instructors engaged students in physical explorations of math concepts, grounding abstract concepts in concrete experiences. To explore symmetry, the instructors first modeled movements and invited the class to categorize the movements and to determine if they were symmetrical or asymmetrical. These physical experiences later served as reference points in students' digital design projects. While working on designing symmetrical dances in the Choreo Graph app, two students tried to determine whether the movements of their animated characters were symmetrical by physically acting out the movements with their bodies and pointing out that "most of the body is moving over *here*, but what about over here?"

Many times, the room fell quiet as every student focused on the projects they were working on, starting with some of the same parameters yet following widely varied trajectories. Students that met the first set of parameters of a particular problem would often be given an additional challenge, such as moving from creating a single-figure symmetrical dance to creating a double-figure symmetrical dance. Sometimes the engagement was raucous, as students dashed around NYSCI's Science Playground in search of angle examples in their everyday life or danced to salsa music while making angles with their bodies. Students who created symmetrical photos laughed and debated their poses.

The confidence the students gained by presenting in front of the group was evident by the end of the week. One student said:

> My favorite part of this week was how I was able to get confident and trust myself and be able to, like, not get scared of how people were around me and not get shy when people try to talk to me. I am like that person who when you start talking to me, I start shaking.

This same student noted that before this workshop, she would typically refuse to present in school, even if the teacher asked her to. When the interviewer asked her whether, after this workshop, she would be able to present in class, the student said she probably could.

dorothy bennett et al.

Finally, students talked about how they had learned math and had fun at the same time, which they said did not often happen in school because activities were repetitive, or teachers did not always have time go over things they missed the first time around. This was true for concepts like fractions that many were already familiar with, and for concepts like symmetry that they were learning for the first time. Overall, students had experienced a different path into mathematics learning and gained confidence in sharing what they knew and could do.

ELL Teachers' Journey from Formal to Informal Instruction

While it was a powerful experience for students to be involved in the workshops, the workshop experience also had an impact on the lead ELL teachers and the teacher-observers.

Our two lead ELL teachers bravely navigated the journey from formal classroom instruction to informal instruction using new tools and approaches. We asked them to not only learn new digital tools rather quickly but to codesign the program of activities and to lead sessions with students every day of the weeklong workshops. In kicking off the first pilot workshop session, they announced to the students, "Remember we are here to have fun – FUN-FUN-FUN!" as a reminder that these sessions would be different from formal classroom teaching of the required curriculum. Several aspects of the pilot's design, discussed below, helped to make this journey feasible and impactful for the lead teachers.

Negotiating Roles

From the start, the lead ELL teachers expressed some confusion about their roles and what the children were expected to learn from the workshops. It was hard for them to take off their classroom teaching hats for a bit and adopt a looser approach. In interviews conducted immediately after the workshop, they spoke of this tension. The TESOL educator was surprised to discover that students came into the workshop with misconceptions about fractions and angles, two topics that would have been covered rather early in the curriculum. Both teachers spoke of how the workshop format allowed them to see those misconceptions in new ways. At the same time, the teachers were surprised that engaging students in designing with these tools readily revealed students' strengths and skills that otherwise would not have been apparent. The teachers also appreciated the possibilities for differentiation that the workshop approach provided. The tools offered multiple points of entry for students to engage with mathematics at their own level.

Team-Teaching

Our lead ELL teachers, who had never met before, were surprised at how well they complemented each other. They bonded quickly and bounced ideas off each other about how to proceed during each day of the workshop. The encouragement from NYSCI staff to be playful and to iteratively codevelop activities as they went along made them feel more prepared and comfortable in the informal environment. The TESOL educator valued the opportunities to alter and reflect on activities, which are not always available in a formal education setting:

> A lot of times with schools we don't ever have time to talk about what worked and what didn't work. It's just like, "Oh, that didn't work. Toss it all out," whereas if we just sat and talked as a team about what worked and what didn't work, better things would happen. And I think we did that this time.

Tapping Parental Input

Our lead teachers viewed the inclusion of parents and family members as a big plus because it enabled families to celebrate what the students had accomplished in a short period. In addition, the positive feedback that teachers received from parents at pick-up and drop-off times helped them get to know the students better and understand their unique sociocultural contexts.

Power of Informal Learning Through Design and Play for ELL Teachers

Three years later, both lead ELL teachers continue to speak of how much they got out of working in an informal environment that emphasizes giving children time to play. The TESOL educator explained the benefits in this way:

> It was a challenge for me at first, thinking outside the box of formal education, but the process was fluid and I loved it. I now try to bring this into my own teaching and have my own maker space for writing, and I allow much more time for play with numbers in math and see the value of the discussion work in the project. This has improved my own teaching.

The bilingual educator found that the project encouraged her to use technology more fluidly in her teaching and inspired her to be creative in her own work. This has been especially helpful during the COVID-19 pandemic, as much of her teaching migrated online and she had to create engaging entry points for students to communicate with her in an online environment. She credits this work with also opening doors to becoming a professor at two colleges, where she helps to teach other educators how to support ELL students. She came to realize that all of the workshop designers were reimagining education through their early adoption of these open-ended tools. But she also stressed that it was not just the technology – she is convinced that activities such as sending children on angle hunts throughout the museum probably changed how children see the world. "They probably can no longer avoid seeing angles or pentagons in the things around them," she noted. "The biggest joy in this work was becoming real facilitators and learning how important it is for children to explore and play with ideas."

dorothy bennett et al.

Broadening Teacher-Observers' Views of What Mathematics Learning Looks Like

As explained above, four additional teachers spent one or two days at the first pilot workshop observing their students, recording instances of mathematical "talk," and thinking about what they witnessed. The goals were to enable teacher-observers to see what their students were capable of in an informal mathematics context and to broaden their conceptions of mathematics discourse among students whose first language is not English. Initially these teachers were unsure what they would gain from observing their students in our setting, but they ended up having unexpected insights.

When they began observing students, some of the teachers questioned how they might apply the teaching approaches used during the workshop to their own classrooms, where resources were scarce. In addition, they found it difficult to step back and observe without intervening to "teach." By the end of the sessions, however, all of the teachers expressed surprise at what their students had demonstrated in the workshop context and felt it gave them a new window into their students' strengths and challenges. One teacher noted the change in the confidence of a shy student over the course of the first day, something she had not seen before in her own classroom:

> What surprised me was her "aha" moment towards the end of the day, more so the look of fulfillment on her face and the way [she] popped out of her seat to command the instructor's attention. It's the physical act that surprised me more; as mentioned, she started off very quiet and reserved.

Conclusion

From the pilot workshops that blended formal and informal learning, several themes emerged with implications for designing future programs to foster inclusion and agency for ELLs. It was clear from our teacher partners that our playful approach and design-based strategies, as well as our efforts to create a judgment-free zone, were powerful ways to help children feel safe to experiment, build confidence, and reveal their strengths and their challenges. For our lead teachers, inviting them to actively play and observe rather than "teach" freed them up to see students and their needs in a new light and to reflect on and rejuvenate their own teaching approaches. As for our museum team, we learned valuable lessons about the community of learners who often attend our museum. We also developed a renewed commitment to collaborate with teachers to create multimodal design activities that can support ELLs' learning in and out of school.

At the same time, there were natural tensions in striking the right balance between formal teaching and free play. What made this collaboration work is that early on we acknowledged that we were going to iteratively revise our activities using existing resources and that teachers, museum educators, researchers, and the youth helping us had genuinely relevant and important skill sets and perspectives that were necessary to make this work. The entire team recognized that these resources were not readily available in the formal classroom or in the informal context alone. This not only built trust but focused us on the shared goal of working toward one thing – ensuring that students had fun with mathematics and engaged in discourse to express themselves.

Looking forward, this project work suggests that informal institutions can play a critical role in becoming informal labs, judgment-free zones, or third spaces where ELLs can build valuable skills at their own pace and where teachers can get to know their students better and understand better which strategies work best for reaching them. Fostering agency and inclusion is dependent on how one allows children to reveal their strengths and what support is provided to make that happen. For English language learners in our program, this process meant enabling them to explore mathematics with all of their senses and express themselves in a variety of modes with formal and informal educators who were willing to play right alongside them.

Acknowledgments

Funding for this pilot program was provided by the Verizon Foundation. NYSCI's Noticing Tools apps, which were central to the program, were funded by the Bill and Melinda Gates Foundation.

References

Garcia, E., & Gonzalez, R. (1995). Issues in systemic reform for culturally and linguistically diverse students. *Teachers College Record, 96*(3), 418–431.

Hindley, L. M. (2003). Reactions of LEP (Spanish) students to four methodological approaches in a 9th grade mathematics class. (UMI No. 3091255) [Doctoral dissertation, Columbia University]. ProQuest Dissertations Publishing.

Lee, V. (Ed.). (2015). *Learning technologies and the body: Integration and implementation in formal and informal learning environments.* Routledge.

Moschkovich, J. (2012). Principles for mathematics instruction for ELLs. Understanding language: Language, literacy, and learning in the content areas. https://ell.stanford.edu/sites/default/files/math_learnmore_files/2.Principles%20for%20Math%20Instruction%208-14-13.pdf

Quinn, H., Cheuk, T., & Castellón, M. (2012, April). *Challenges and opportunities for language learning in the context of Common Core State Standards and Next Generation Science standards* [Conference overview paper]. Stanford University Understanding Language Conference. http://ell.stanford.edu/sites/default/files/Conference%20Summary_0.pdf

Solnit, R., & Jelly-Schapiro, J. (2016). *Nonstop metropolis: A New York City atlas.* University of California Press.

chapter 10

Learning Physics through Embodied Play in a School Setting

Harouna Ba, Christina O'Malley,
Yessenia Argudo, and Laycca Umer

How students were able to communicate their understanding gave me chills!!! This is physics that I struggled with when I was in high school, yet these middle schoolers made the connections without a minute of direct instruction!!! After giving students three class periods to explore the playground physics app, kids made connections between motion, distance, time, speed velocity, and acceleration and then extended connections to Newton's laws and forces.

Teacher in the case study

Can I take my Chromebook with me snowboarding this weekend? I want to use Playground Physics to check out my jumps.

A student participating in Playground Physics

A hallmark of informal science learning environments is "embodied" learning, in which participants actively engage with tasks, materials, and interactive exhibits (Honey & Kanter, 2013; Lee, 2015; National Research Council [NRC], 2009). In these informal settings, embodied activities are often playful and involve using one's hands or whole body. They are designed to stimulate high levels of agency and autonomy through social and playful interactions, multifaceted and dynamic portrayals of scientific information, opportunities to actively shape tasks and materials, and inquiry and discovery processes. Research has shown that features of informal learning settings "can support an openness, readiness, and willingness to try new things and work with peers" (Bevan & Michalchik, 2013, p. 203).

This approach to engagement and learning is often absent from instruction in science, technology, engineering, and mathematics (STEM) in formal learning environments. Schools tend to treat STEM instruction as "serious" business. Among the biggest challenges teachers face in middle schools are student apathy, lack of motivation, and disengagement (Quinn & Cooc, 2015; Wang & Holcombe, 2010), especially in science classes (Lyon et al., 2012). Science curriculum activities are often prescriptive and rigid. They make little room for personal investment and are focused on drill-and-skill and memorization rather than problem-solving. This approach not only leads to high rates of student boredom, alienation, and low science achievement, but also contributes to persistent challenges such as high dropout rates (Fredricks et al., 2016), widening social stratification, and STEM employment gaps (Education Commission of the States, 2018). Disengagement from science is most pronounced among middle school students, particularly females and students from underserved groups who have few opportunities to engage critically and constructively with complex science concepts (Britner & Pajares, 2006; Lyon et al., 2012; National Academy of Sciences, 2011; NRC, 2008).

A key question, therefore, is how educators can leverage the intrinsically motivating and playful embodied activities of informal learning environments to develop innovative science curricula, professional development, and digital tools for formal STEM learning environments.

At the New York Hall of Science (NYSCI), we are exploring ways to leverage the most effective aspects of informal learning environments to support students' productive engagement with core physics concepts in formal learning settings. By "productive engagement" we mean physical play, such as running, jumping, and sliding, and investigations such as using an app to capture and explore these play performances through a physics lens. In our work at NYSCI we use the Design Make Play pedagogical approach, which puts the learner at the center of the learning process, capitalizes on learners' inclinations to engage playfully with things they find compelling, and offers problem-solving opportunities driven by their interest and experiences. (See this book's Introduction for more details about the Design Make Play principles.)

This chapter focuses on implementation of and insights from Playground Physics, a program for middle school teachers and students designed by NYSCI using the Design Make Play approach. Playground Physics integrates both informal (playground, hallway, gym) and formal (classroom, science lab) activities to improve productive engagement with core physics concepts among middle school students from underrepresented groups. Playground Physics consists of a digital app and curriculum activities that enable middle school teachers to implement the program in their classrooms and help students connect complex physics concepts to what they do in real life, including their deep, embodied intuitions about the concepts of motion, force, and energy. It invites students to reason and discuss their experience and fosters rigorous academic physics learning (Friedman et al., 2017; Margolin et al., 2020).

Since Playground Physics combines the engagement qualities of informal learning environments with rigorous academic learning, we are using our experiences with this program as a case study to explore how teachers and students navigate these opportunities in school environments. We partnered with a science teacher to investigate the first phase implementation of the program, which familiarizes students with the app, supports them in asking and investigating their own questions, and encourages them to use their own video-recorded play as data for making sense of their experiences. We looked into the implementation of this first phase of the program, students' engagement and disengagement, and effective strategies for supporting engagement across school spaces like classrooms and hallways.

Research Evidence for Our Approach

The design of Playground Physics was informed by evidence on two key aspects of the program: embodied play and playfulness, and digital tools.

Embodied Play and Playfulness

Play and playfulness are compelling frames and strategies through which children explore and notice the world around them (Bergen, 2009; Singer et al., 2006). They help children combine thoughts or actions in innovative ways, which can lead them to think divergently and come up with novel solutions to complex problems (Bateson & Martin, 2013). These types of thinking and practices reflect, albeit at a more basic level, how scientists approach complex scientific problems (Bergen, 2009).

Embodied play and playfulness offer a means to anchor the learning of complex scientific concepts by engaging students' senses and encouraging them to investigate representations or models (Abrahamson & Bakker, 2016; Galetzka, 2017; Lee, 2015; Lindgren, 2014; Trudeau & Dixon, 2007). For example, researchers have used embodied design problems to support student discussions about and reflections on complex scientific concepts. Examples of these problems include making a molecule (Enyedy & Danish, 2015), following a meteor (Lindgren, 2014), modeling bees (Danish et al., 2018), and moving in uniform angular velocity (Zohar et al., 2017). In informal science learning environments, embodied play and playfulness are quintessential ingredients that enable visitors to actively explore complex scientific ideas, multifaceted and dynamic portrayals of science data, and interactions with scientific phenomena (Association of Children's Museums, 2017; Lee, 2015; Peppler, 2017; American Institutes for Research [AIR], 2016). Through interacting with skilled facilitators who ask relevant questions and provide guidance, learners can actively shape the tasks they are working on. This approach has been shown to positively impact students' motivation and engagement (AIR, 2016) and enrich learning (Robertson et al., 2015).

Embodied play and playfulness are at the core of the Playground Physics program. In Playground Physics, the embodied design problems are grounded in physical play such as running, jumping, and sliding; in open-ended design problems; and in child-generated movement problems such as how to generate maximal potential energy or kinetic energy. These problems are mediated by technology and investigated with instructional support (Ba & Abrahamson, in press).

Digital Tools

Research has shown that digital resources enhance engagement with and motivation to learn science in ways that have few real-life referents (Anderson & Barnett, 2014; D'Angelo et al., 2014; Clark et al., 2016). Digital tools can be used to support science instruction (Clark et al., 2016) and have been found to be effective when used in conjunction with other experiences that introduce students to science concepts (Smetana & Bell, 2012). For example, students using digital tools such as simulations and games can see and interact with representations of natural phenomena that would otherwise be difficult to observe (Rutten et al., 2012; Honey & Hilton, 2011). In studies of physics learning, researchers have shown that well-designed simulations and games have a positive impact on students' scientific understanding (Shute et al., 2013; Squire et al., 2004). In the same vein, cognitive scientists are integrating embodied experiences with digital resources to promote STEM engagement and learning (Abrahamson, 2014; Anderson & Barnett, 2014; Lee, 2015). Digital resources can serve as mediating resources (e.g., diagrams, inscriptional overlays) that enable users to fuse videos and graphs to generate referents and symbols in the learning space (Enyedy et al., 2012; Nemirovsky et al., 1998).

These types of learning experiences are available in the Playground Physics app. The app uses real, not simulated, data from students' embodied performances on a playground to investigate their own spontaneous play in a digital environment. Playground Physics is not a game; the external motivators that researchers mention for games, such as instant gratification, wins, and points (Deterding et al., 2011), are not the goal of students' interactions with Playground Physics. Instead, the app invites students to explore their own play activities. This can help them to notice something about the relationship between their spontaneous embodied experiences, their prior physics knowledge, and the scientific ideas embedded in a curriculum.

From a Science Playground to a School Physics Program

The Playground Physics instructional program was inspired by the quality of active engagement taking place in NYSCI's outdoor Science Playground exhibit. Playground Physics is intended to bring the playful, engaging experience of Science Playground to the more formal learning environment found in schools.

Science Playground Exhibit

On any day of the week during spring, summer, and fall, visitors of all ages are actively engaged in embodied play and playful experiences at NYSCI's Science Playground, a 60,000-square-foot outdoor area. The playground environment, which is equipped with dozens of elements like slides, seesaws, sand pits, and fog machines, promotes active engagement, full-body experiences, and multiple modes of open exploration (see Figure 10.1). The playground's safe and challenging learning environment and basic physics principles encourage visitors to experiment and make connections to their own experiences. They explore scientific concepts and phenomena such as motion, balance, sound, sight, and simple machines, as well as sun, wind, and water (Chermayeff et al., 2010).

Playground Physics Program

In addition to being grounded in theoretical and empirical approaches to embodied learning, education technology, and informal science learning, the pedagogy of Playground Physics reflects the museum's Design Make Play principles. These principles call for putting the learner at the center of the learning process and emphasizing open-ended exploration, imaginative learning, personal relevance, deep

Figure 10.1 NYSCI's Science Playground.

engagement, and delight. As part of NYSCI's educational strategy of broadening access to quality STEM education in formal learning environments, the program uses the playground as a laboratory for experiencing physics concepts such as motion, force, and energy. Its goal is to make physics learning accessible and fun for middle school students, especially those who have shown little interest in science learning, find it to be challenging and boring, and do not identify with the field of science. Playground Physics combines the engagement qualities of informal learning environments with rigorous academic learning. The program offers students multiple playful ways (physical, emotional, and cognitive) and dynamic portrayals of physics data to explore physics concepts and ground their learning about motion, force, and energy.

Playground Physics consists of an app, supplemental curriculum, professional development activities, and ongoing support for teachers. Installed on an iPad or Chromebook tablet, the app allows students to video-record their full-bodied physical play activities (see Figure 10.2). Students select one body part (a head, leg, hand, or foot) and place dots to trace the path of the selected body part. They enter required calibration information (mass, height, and size) about the selected body part into the app and name and save the video on the tablet. Using the video and other data that students entered and saved, the app generates graphs linked to the video. The app allows students to simultaneously explore everyday physical experiences and physics

Figure 10.2 App and users.

concepts within "one consistent space using talk, gesture, and representations without distinguishing between them" (Enyedy et al., 2012, p. 352). For example, they can see where they are moving the fastest or slowest, where a force is pushing or pulling, and where their kinetic and potential energy are at their highest and lowest points. Students can use digital stickers to annotate these moments.

A six-week supplemental curriculum accompanies the app and is designed to support teachers' implementation of the program in formal classroom environments and informal learning environments within school settings, such as a playground, hallway, or gym. The curriculum, which includes a teacher guide and student workbook, provides three-dimensional instructional materials that authentically engage students with science practices, disciplinary core ideas, and crosscutting concepts. The curriculum highlights the principles of physics that are present in different types of physical play experiences (swinging, sliding, and running, for example) and play performances in the app.

The three units in the curriculum – motion, force, and energy – can be taught in any order. Each unit includes a review of content knowledge as well as common student misconceptions about the topic. Exploration centers on students' play activities, such as playing catch for motion, jumping rope for force, and swinging for energy. The units are presented to students in the form of a guided inquiry and start with a lesson to help students begin thinking about and reflecting on their own experiences. Subsequent lessons build on one another to actively engage students through use of the app and unit activities. The curriculum and app blend embodied physical and digital activities in playful ways to support students' reasoning and discussion of their experiences through the three "lenses" of motion, force, and energy.

The Playground Physics professional development activities are designed to help teachers learn how to use the app, experience the physics content as learners would through playful hands-on activities, and consider the instructional materials as educators. Then, teachers are invited to reflect on the underlying embodied learning approach and think about how to realistically implement the curriculum in their classrooms.

In the classroom, educators receive continued support for Playground Physics through an online community of practice. Online conversations might focus on modifications to the curriculum, scaffolds to move students toward greater understanding and independent learning, examples of student work, or problems of practice. In both the workshops and online community, educators are given agency over how they enact the instructional materials to best suit their implementation contexts. By providing educators with the opportunity to experience aspects of the program as learners (Mundry & Dunne, 2003), the program design intentionally equips them to make these modifications while still staying true to the underlying playful approach to physics teaching and learning.

Connecting Research and Practice

Museum science instructors and researchers worked together and with teachers and students to conceptualize and design the Playground Physics program. The program's implementation strategies were tested in a variety of middle schools and refined over time. During the conceptualization and development of the program in New York City (2011–2016), NYSCI worked with urban middle schools to better understand the assets and constraints of formal learning environments and create a program that will be responsive to the realities of school. We also sought to determine and design the types of ongoing support teachers will need to implement the program in their classrooms, especially in managing the program's technology and embodied play components.

During a new expansion of Playground Physics across New York State (2018–2022), museum educators and researchers are working with instructional technology and science integration coaches and teachers to broaden access to the program across urban, suburban, and rural schools. This phase of collaboration aims to further refine professional development to immerse teachers in the embodied, playful, student-driven learning approach. A second goal of this expansion involves revising curriculum activities to ensure they are rigorously aligned to state standards while maintaining play and playfulness as the core experiences. A third goal is to provide ongoing coaching and support to teachers and promote teacher dialogue through the online community of practice.

As discussed in the next section, we also partnered with one of the teachers participating in the initial implementation of Playground Physics across New York State to look at how she integrated informal and formal learning strategies into her science teaching, how her students experienced engagement and disengagement, and what implementation strategies she used to support productive engagement in physics learning through playful embodied investigations in school spaces.

Case Study Design and Implementation

We used the Playground Physics program as a case study to gain preliminary insights about the implementation of the first phase of the program. In addition, the case study examined students' emotional, behavioral, and cognitive engagement and effective strategies for leveraging and sustaining engagement in a suburban middle school in New York.

As we implemented the first phase of Playground Physics in a school setting, we operationalized our concept of productive engagement through playful embodied investigations by drawing on research in informal and formal learning environments. In school settings, engagement is defined, designed, and supported in the context of learning content knowledge and thus is directly linked to a drive to learn and academic achievement (Bircan & Sungur, 2016; Lee et al., 2016). Through this lens, disengagement is characterized by passivity, boredom, inattention, and related behaviors (Skinner et al., 2009). Research in schools posits three core dimensions of engagement:

- *Emotional engagement*, which focuses on affective reactions such as excitement, interest in learning a particular subject, and a sense of belonging in school settings (Finn, 1989; Ladd et al., 2009; Renninger & Bachrach, 2015; Voelkl, 1997)
- *Behavioral engagement*, which pertains to students exhibiting active, constructive and collaborative participation, adopting classroom norms, paying attention

(concentrating), being persistent and responsive, taking initiative, and not engaging in disruptive behavior (Ladd et al., 2009)

- *Cognitive engagement*, which refers to the degree of investment in the learning activities, the level of processing or intellectual effort devoted to specific learning tasks, and the exertion of effort needed to understand and master the material (Fredricks et al., 2016; Ladd et al., 2009)

We drew on these three well-established dimensions of school engagement to inform the design of our observation protocol. We further complemented them with findings about engagement in informal learning environments.

In informal settings, engagement is often encouraged by a set of deeply playful and active experiences that engender curiosity and interest in a topic (Shernoff & Vandell, 2007) and self-efficacy (Bell et al., 2019). These learning environments are often committed to the emotional, behavioral, and cognitive participation of visitors (NRC, 2009). They build on visitors' diverse interests and curiosity and support visitors' self-motivated inquiries (Friedman, 2003; Humphrey et al., 2005; Lemke et al., 2015). We believe that in informal learning environments, engagement is related to enjoyment and agency. Although definitions of engagement vary in informal settings, engagement has been defined as "affective involvement in and commitment to an activity, goal, practice . . . that enhances the quality and quantity of participation despite obstacles, setbacks, or frustrations." Enjoyment has been defined as the "positive feeling accompanying an activity that makes it worth doing for its own sake," while agency is related to "actual effectiveness, a disposition toward taking action, a feeling of self-efficacy, and an aspect of one's identity as someone who can produce desired effects" (Lemke et al., 2015, p. 12).

To develop an observation protocol, we used these definitions, as well as existing standardized classroom and playground observation research instruments designed to measure engagement (Birdwell et al., 2016; D'Mello et al., 2017; Cornelli Sanderson, 2010; Engelen et al., 2018; Fredricks et al., 2011; McKenzie, 2006; Renninger & Bachrach, 2015). The protocol has three sections: (a) setting, environment, and participant information; (b) emotional, behavioral, and cognitive engagement items during preparation time in the classroom; and (c) an in-depth checklist designed to document students' engagement during playing, filming, and debriefing times. The observers have access to open-ended subsections of the observation protocol where they can note unexpected behaviors and actions demonstrating engagement.

To conduct the case study, we worked with one science teacher and her 59 students at her school. Each of her five science classes has on average 12 students. We observed five 45-minute class periods. Each period was divided into 15 minutes for introducing the day's activities, 20 minutes for playing and filming, and 5 minutes for debriefing. We documented a total of 59 types of emotional, behavioral, and cognitive engagement across three settings (classroom, computer lab, and hallway) and three key times (prep time, playing and filming time, and debriefing time). After these observations were completed, we asked the teacher to share her implementation experience, observations of student engagement, and reflections on the challenges encountered during implementation. She described how she prepared and implemented the lesson, shared the questions she asked her students and a summary of their responses, reflected on the overall implementation of this program phase, and provided recommendations for practitioners.

Our preliminary findings from this case study are presented below.

Implementation of the Program in School

The first phase of implementation is intended to promote students' productive engagement for physics learning. Students are introduced to the app, engage in physical play performances, use the app to document those play performances, and review and discuss the quality of their videos for physics learning. While teachers frame and set up the activities, students define their own tasks, generate their own questions about their play performances, and make their own choices about how they want to document their play performances. Back in the classroom, they work collaboratively in small groups to share and discuss their embodied play experiences without referencing the app. They investigate their embodied performances using the videos through the motion, force, and energy lenses and compare what they felt against the data provided by the app about how fast or slow they were going, how high they were able to swing, and more. Students' rich intuitions about the concepts of motion, force, and energy are made visible through their physical play performances, which allow them to reflect on and question their intuitions (Enyedy et al., 2012). This process is a continuous cycle of discovery and improvement: students test and retest their play performances and videos in order to answer their movement questions about what aspects of their play performances make a difference in physics learning. Guided by the curriculum and professional development activities, teachers play a central role in ensuring the effective implementation of this first phase of Playground Physics.

This first phase of implementation in five science classrooms went through three stages as the teacher coached students about how to feel comfortable with playing in a formal learning environment while keeping a respectful noise level and how to use the app to get optimal recordings for physics learning.

Stage 1

Students self-selected into teams of three or four and agreed to roles in their teams, such as main camera, director/camera back up, actor 1, and actor 2/set prep. The teacher reviewed the student workbook sections on "how to use the app" and "tips for filming" and addressed students' questions about technology use related to their Chromebooks. The teacher then walked students through capturing and annotating a video by projecting her Chromebook onto a smartboard. The teacher introduced sports equipment, props, and toys that she had set up around the room. Students were instructed to film themselves doing any playful activity they would like to do, following along with information in the workbook to learn how to record, create a path for performance, enter size, and watch their performance. Students could freely move between playing and filming in the classroom, lab, and hallway until they were called back to discuss their experiences with the app and their suggestions for making filming easier.

Stage 2

Students gathered materials and spread out to explore the app. They were encouraged to practice fearless exploration: to play around with filming in order to learn how to use the app and discover the physics that goes on during their physical play. Above all, they were encouraged to have fun while respecting the other classes occurring at the same time. Students expressed excitement and discussed how they liked to use the app to record and explore "running," "checking how fast I can run and my progress," measuring "how fast I can kick a soccer ball," "see[ing] the different activities we do and how it affects our motion and energy," and "watching a race . . . through a different viewpoint . . . and seeing the motion that is behind each move/play." They took advantage of the opportunity to work inside and outside the classroom and engaged in a variety of play and filming activities. Some groups of students passed a hockey puck back and forth, others tossed a football, a few groups did cartwheels, and others slid their bodies across the floor, among many other activities. Students discussed their questions, opinions, and suggestions about the activity of playing and recording. A few students tried to record the action vertically rather than horizontally, and some forgot to set up a yardstick. Many did not get their Chromebook in a location to capture all of the action in the scene. The teacher provided facilitation and guiding questions, such as suggesting that students measure against a tile on the wall or the height of a team member in the background in lieu of using a yardstick.

Stage 3

Students gave back the props and returned to their seats. They shared important tips they had learned about how the app works, specifically how to make a successful recording, and verbalized ideas for how they may want to use the app the next time. When asked how they traced a path in a performance, they shared different strategies about what to select as the focal point of their performance, as illustrated by these comments:

- *Mark the thing that is present throughout the whole motion.*
- *Go frame by frame placing dots.*
- *Pick the most active parts of the body.*

Their explanations of how they used the app to determine speed in a performance ranged from "the app records it for you," to "[I] pulled up the speedometer," to "you used the dots to mark where the 'actor' went and it calculates the speed." When asked how they determined height in a performance, they offered the following answers:

- *Had a measurable object in the background*
- *Held a yardstick in the shot*
- *Used a yardstick and measure[d] the height and then [put] the height into the app*

Implementation Challenges

Despite her successful implementation of this first phase of the program, the teacher reported two key implementation challenges: lack of time, and tension between open exploratory learning and directed instruction. The first challenge involved finding time during a 45-minute class period to negotiate how to best support students as they moved from space to space, engaged in play activities, used the app to record their activities, collaborated with peers, and reflected on their experiences. The second challenge related to problems students experienced in operating the app. Although the teacher appreciated the format of the curriculum, which keeps exploratory learning in the forefront, she observed that students did not successfully learn all the skills needed to operate the app in the first round of playing and filming. For example, she reported that many students forgot to put the full yardstick in the recording, and some struggled to get the whole action in the shot. As a result, they were unable to collect optimal embodied data that could be used during debriefing time to explore in depth the physics concepts of motion, force, and energy and investigate the connections between their playful embodied experiences and these physics concepts. To help students over these types of problems, the teacher had to intervene by providing more direct instruction, which she felt went against the exploratory approach called for by the curriculum.

Evidence of Engagement

As mentioned above, our case study gathered evidence about three core dimensions of engagement: emotional, behavioral, and cognitive.

Emotional Engagement

Emotional engagement was prime in the implementation of the first phase of Playground Physics. More students evidenced emotional engagement (59%) than either behavioral (27%) or cognitive (14%) engagement. We observed them exhibiting four out of five indicators of emotional engagement: they were excited, interested, having fun, and happy across activities and spaces. (The fifth indicator, pride, was less apparent.)

Emotional engagement was more prominent during prep time than during the other two times. The first time students were shown what the Playground Physics app could do, their excitement was visible and audible, even though this occurred during a time when they were often expected to be silent and listen to the teacher's instruction. Although students' overall emotional engagement was less evident during playing and filming time, one of the five indicators of emotional engagement – having fun – increased almost fourfold from prep time (7% of students) to play performances and filming time (27%). We observed students smiling, laughing, cheering, and acting silly during recording and playing time.

Behavioral Engagement

Behavioral engagement was observed using four indicators: participating, comparing and discussing, collaborating, and seeking support. We observed 27% of students

harouna ba et al.

physically playing, actively collaborating, and seeking each other's support. During prep time, most of the students (67%) looked really involved in the activities, and 33% collaborated through listening and interacting with each other in supportive ways. We observed one group of students as they struggled to transition from being rowdy and impatient during prep time to preparing for their play and recording activities. They worked collaboratively to discuss and decide each member's specific role during the filming and which activity they would film first.

During playing and filming time, we observed 40% of the students playing harmoniously with others, 30% looking really involved in the activities, and 30% collaborating through listening and interacting with each other in supportive ways. Students also seemed quite involved in experimenting with a variety of play performances and video recordings. In one of the classes, we observed a group of students working together to review the videos, apply stickers, discuss measurements, and generate feedback on their play performances.

During debriefing time, most students (67%) worked with other students and learned from each other, and 33% looked really involved in the activities. We observed students talking, reviewing their videos, and showing the functionalities of the app. They also appeared to actively interact with the teacher and answer the teacher's questions.

Cognitive Engagement

Cognitive engagement was determined using three indicators: asking questions, sharing ideas, and participating in discussions. We observed 14% of students sharing ideas and suggestions for how to best perform and shoot their play performances using the Playground Physics app. Of the 14% of students who exhibited this type of sharing, most (75%) did so during play and filming time. Specifically, these students were observed thinking about how to best perform their play activities and shoot their play performances. They were seen providing suggestions about issues of video framing, field view, distance, angles, and tablet stability during the performance and filming.

During debriefing time, 67% of students worked with other students and seemed to learn from each other, and 33% appeared to be really involved in the discussion activities. They interacted with each other and actively discussed their experiences in using the app and filming with other students. They talked, reviewed their videos, and showed the functionalities of the app. They also interacted with the teacher and their peers as they discussed their understanding of how the app works and how to engage in play performances and make videos for physics learning. They reflected on lessons learned about how to engage in playful embodied physics learning, as in these student comments:

- *I want to see what treat makes my dog run faster.*
- *How does changing gears of my dirt bike change my acceleration?*

A participating teacher described the student discussions and benefits of the program for physics learning:

> I could hear students formulating ideas for what they would like to record and get data on; there was definite excitement going on! . . . The format allows students

. . . to explore to establish prior knowledge in the functionality of the technology, so that they may move forward to focus on the physics connections.

Curiosity as a Driver of Engagement

Although students struggled to record useful data for physics learning on their initial attempts, the teacher indicated that students' curiosity prompted their efforts to plot points, add height, and observe their recordings through the lenses of the three physics concepts. Nearly all groups of students in her five classrooms were starting to explore how the app collects and generates data and to grapple with concepts of speed and height on their own accord. When prompted by the teacher during debriefing time, at least three-quarters of students verbally shared their preliminary understandings of the differences between the motion, force, and energy lenses in the app. All students demonstrated that they had gained some understanding of how the app works and some of the data generated by the lenses, as evidenced by students' explanations like these:

- *Motion compares speed, energy compares potential and kinetic energy.*
- *Force is push or pull, motion is moving, energy is living.*
- *They all have different types of units to them. It [the app] tracks the force, the speed, and the potential and kinetic energy.*

Students were also asked to imagine how a scientist might use the lenses to conduct scientific investigations. They suggested that experts can use the app to measure speed, force, and time; identify the location of scientific events; and explain the behavior of objects in the different lenses.

Predefined Roles as a Driver of Disengagement

The teacher reported on one case of student disengagement related to roles and responsibilities during play and filming. She observed that having teams decide on roles before engaging in the task was demotivating. As this phase of the program rolled out, many of the groups with predefined roles continued to display less enthusiasm and placed the responsibility of plotting points and analyzing data heavily on the student in the main camera role. She explained that the classes that did not establish student roles before engaging in playing and filming were more engaged when exploring for the first time. They focused more on making fun recordings and were able to organically share and rotate roles.

Productive Engagement Strategies for Supporting Physics Learning

As described above, the implementation of the first phase of the Playground Physics program presented instructional and logistical implementation challenges for both students and teachers. During a 45-minute period, many students had problems making videos with the app and were unable to collect the quality of data needed to investigate in depth the connections between their playful embodied experiences and the concepts of motion, force, and energy during debriefing time. Although the teacher in our case study acknowledged the power of choice, agency, embodied play, and playfulness in engaging students in physics investigations, she struggled to determine how much direct instruction and guidance she should provide without dampening the high level of emotional and behavioral engagement shown by her students during playing and filming times.

The teacher indicated that the following strategies were effective in helping her deal with these challenges of leveraging student engagement to investigate physics concepts:

- *Integrate both open and guided approaches to playful exploratory learning and rigorous academic instruction in balanced and supportive ways* (Wesiberg et al., 2013; Fine, 2014). This type of integration is critical as students play, learn to master digital tools, and use these tools to collect data to attempt to answer their initial intuitive questions about their embodied experiences. Open playful exploration is generally framed by the teacher but is fueled by student agency and engagement in informal learning activities. Guided playful exploration draws on student agency but is teacher-led; the teacher provides instructional guidance geared toward formal academic learning. Both open and guided exploration require a shift in practice whereby the teacher relinquishes total control of the classroom so that activities are student-directed and promote student engagement and agency.

  A basic place to start with Playground Physics is to allow students to self-select or naturally take on specific roles during playing and filming time, such as performer, recorder, or producer, rather than pre-assigning roles. This is critical because this is the point at which students are most likely to exercise their agency over the activities and be intrinsically motivated to engage with peers to explore the connections between embodied play and physics learning.

- *Use formative assessment throughout this phase.* Formative assessment has many useful functions, such as monitoring students' familiarity and competence with the content topic and/or digital tools, gauging students' interest in the content and activities, and gathering student feedback to help frame future instruction. It is critical to provide immediate feedback to students so they can improve their experiences and understandings (Keeley, 2016). These types of formative assessments can be done through a short online survey, such as a Google form, used as a brief recall activity. The assessments should be low-stakes and not graded. This low-stakes form of assessment allows students to practice retrieving what they have learned and gives them ownership over their understanding. At the same time, it enables teachers to see where gaps in understanding remain.

  With Playground Physics, it is useful to monitor students' implementation of the activities across environments throughout this phase. Monitoring enables teachers to identify what additional support students need in order to collect optimal data for physics learning and to remedy any problems. For example, the teacher reported that she observed students struggling to make a proper recording, plot a path for data collection, and input the required physics quantities like height and mass. To address this, she decided to demonstrate these skills continually to students over three class periods. To gather feedback, she administered a brief online survey to give students the opportunity to individually reflect on what they knew about the app and what they understood about their experiences. Further, the teacher allowed time for debriefing and questions at the end of each class, during which students discussed what they had noticed, had trouble with, or didn't understand.

- *Expand the time for this first phase of the program to three class periods.* Adjusting it from one class period to three would provide more time for students to understand new concepts and confront their misconceptions at their own pace. This is preferable to frontloading facts about a content topic and expecting students to immediately implement this new knowledge as they use digital tools to collect data or engage in any subsequent instructional activity.

In the case of Playground Physics, for example, the first class period could be used for preparation in which students explore in depth the app's functionalities. The second period could involve playing and filming in order to generate optimal data for learning; these activities could be repeated until students feel that they have produced play performances and videos of the quality needed to address their own questions. The third class period could be devoted to reviewing videos and debriefing about students' play and filming experiences.

Conclusion

This case study provides some insights into the emotional engagement of learners when they are first introduced to playful embodied investigations of physics concepts across learning spaces in a school setting. Curiosity is a key driver of learners' efforts to engage with the tools and activities. In a formal learning environment, the teacher in this case study had to figure out how to effectively manage the tensions inherent in integrating informal and formal learning approaches. Specifically, she sought ways to leverage high levels of student emotional engagement to support their behavioral and cognitive engagement and ultimately set them up for understanding complex physics concepts.

Implementation of the first phase of Playground Physics in a school setting demonstrates that it takes time to productively leverage students' emotional, behavioral, and cognitive engagement and to optimize students' physical play performances and videos for physics investigations. Teachers must ensure that students take on the full responsibilities of determining and assigning roles during playing and documentation. Teachers also need to balance explorative learning and direct instruction and undertake continuous formative assessment. These implementation strategies are aimed at moving from generating productive engagement to supporting it, and ultimately to leveraging that engagement to develop deep understanding of complex STEM concepts.

Furthermore, this case suggests that the design and implementation of this initial phase is a critical mediator between engagement and learning. It should not be minimized or rushed in order to rapidly get to the "serious" business of STEM content learning in the classroom. Play, engagement, and learning are not in conflict in this learning environment. The partnership that emerged among informal educators, practitioners, and researchers throughout the development and expansion of Playground Physics has been critical. This partnership has yielded insights for understanding, designing, and supporting effective approaches to promote student engagement and teacher facilitation of complex physics concepts using technology-enhanced, playful, embodied approaches to instruction. The partnership has also shed light on how to prepare teachers to change their practice to support student-driven inquiry and allow it to thrive.

More partnerships of informal and formal educators and researchers are needed, however, to assist teachers in this transition and improve our understanding of the design and implementation of productive engagement across STEM learning environments. Ideally, future research led by these partnerships will examine effective approaches to providing professional development and ongoing support for productive student engagement. Future research will also address ways to change teacher practice, including preparing teachers to allow students to engage in open-ended exploration, encouraging teachers to relinquish total control of the classroom, and shifting their roles from purveyors of knowledge and facts to facilitators of student-driven play and engagement in scientific content knowledge.

Curriculum Availability and Acknowledgments

The Playground Physics curriculum is available for free on NYSCI's website (www.nysci.org). The iPad app is available for free in the Apple App Store, and the Chromebook Extension is available for free in the Chrome Web Store. The research and development of the program, which started in 2012, has been widely disseminated (Friedman et al., 2017; Kanter et al., 2013; Margolin et al., 2020). It is currently being studied in an expansion initiative across New York State.

The Playground Physics program was supported by Sara Lee Schupf and the Lubin Family Foundation, BNY Mellon, the John D. and Catherine T. MacArthur Foundation, Motorola Solutions Foundation, the National Science Foundation through Award No. 1135202, and the US Department of Education through Award No. U411C110310 and Award No. U411B180028. Any opinions, findings, and conclusions or recommendations expressed in this material are those of the authors and do not necessarily reflect the views of the aforementioned funders. We would also like to thank our team of curriculum and technology designers, researchers, and evaluators who supported the successful development and implementation of the Playground Physics program: Michaela Labriole and Judith Hutton at the New York Hall of Science; Jake Barton and Eric Mika at Local Projects; Trent Oliver, Bryan Wagman, and Robert Moskal at Blue Telescope; Jonathan Saggau at Enharmonic; Amy Perry-DelCorvo and Linda Brandon at the New York State Association for Computers and

Technologies in Education; and Jonathan Margolin, Jingtong Pan, Leah Brown, and Lawrence Friedman at the American Institutes for Research.

References

Abrahamson, D. (2014). Building educational activities for understanding: An elaboration on the embodied–design framework and its epistemic grounds. *International Journal of Child-Computer Interaction, 2*, 1–16.

Abrahamson, D., & Bakker, A. (2016). Making sense of movement in embodied design for mathematics learning. *Cognitive Research: Principles and Implications, 1*(33).

Anderson, J. L., & Barnett, M. (2014). Learning physics with digital game simulations in middle school science. *Journal of Science Education and Technology, 22*(6) 914–926. http://soe.unc.edu/anderson/~anderjl/Middle%20School%20Games%20Final.pdf

Association of Children's Museums. (2017). *Children's Museum Research Network.* https://childrensmuseums.org/childrens-museum-research-network

Ba, H., & Abrahamson, D. (in press). Taking design to task: A dialogue on task-initiation in STEM activities. *E-journal of the International Society for Design and Development in Education.*

Bateson, P., & Martin, P. (2013). *Play, playfulness, creativity and innovation.* Cambridge University Press.

Bell, J., Besley, J., Cannady, M., Crowley, K., Grack Nelson, A., Philips, T., Riedinger, K., & Storksdieck, M. (2019). *The role of engagement in STEM learning and science communication: Reflections on interviews from the field.* Center for Advancement of Informal Science Education. www.informalscience.org/sites/default/files/CAISE%20Engagement%20Overview.pdf

Bergen, D. (2009). Play as the learning medium for future scientists, mathematicians, and engineers. *American Journal of Play, 1*(4), 413–428.

Bevan, B. & Michalchik, V. (2013). Out-of-school time STEM: It's not what you think. In B. Bevan, P. Bell, R. Stevens, & A. Razfar (Eds.), *LOST opportunities: Learning in out-of-school time* (pp. 201–217). Springer.

Bircan, H., & Sungur, S. (2016). The role of motivation and cognitive engagement in science achievement. *Science Education International, 27*(4), 509–529.

Birdwell, T., Roman, T. A., Hammersmith, L., & Jerolimov, D. (2016). Active learning classroom observation tool: A practical tool for classroom observation and instructor reflection in active learning classrooms. *Journal on Centers for Teaching and Learning, 8*, 28–50.

Britner, S. L., & Pajares, F. (2006). Sources of science self-efficacy beliefs of middle school students. *Journal of Research in Science Teaching, 43*(5), 485–499.

Chermayeff, J. C., Blandford, R. J., & Losos, C. M. (2010). Working at play: Informal science education on museum playgrounds. *Curator: The Museum Journal, 44*(1), 47–60.

Clark, D. B., Tanner-Smith, E. E., & Killingsworth, S. S. (2016). Digital games, design, and learning: A systematic review and meta-analysis. *Review of Educational Research, 86*(1), 79–122.

Cornelli Sanderson, R. (2010). Towards a new measure of playfulness: The capacity to fully and freely engage in play [Doctoral dissertation, Loyola University Chicago]. *Dissertations 232.* https://ecommons.luc.edu/luc_diss/232.

D'Angelo, C., Rutstein, D., Harris, C., Haertel, G., Bernard, R., & Brokhovski, E. (2014). *Simulations for STEM learning: Systematic review and meta-analysis.* SRI Education.

Danish, J., Humburg, M., Tu, X., Davis, B., & Georgen, C. (2018). Modelling bees by acting as bees in a mixed reality simulation. In J. Kay & R. Luckin (Eds.), *Rethinking learning in the digital age, 13th International Conference of the Learning Sciences 2018* (pp. 1276–1278). International Society of Learning Sciences.

Deterding, S., Dixon, D., Khaled, R., & Nacke, L. (2011). From game design elements to gamefulness: Defining "gamification." *Proceedings of the 15th International Academic MindTrek Conference: Envisioning future media environments* (pp. 9–15). ACM Press. www.researchgate.net/publication/230854710_From_Game_Design_Elements_to_Gamefulness_Defining_Gamification

D'Mello, S., Dieterle, E., & Duckworth, A. (2017). Advanced, analytic, automated (AAA) measurement of engagement during learning. *Educational Psychology, 52*(2), 104–123.

Education Commission of the States. (2018). *New York trends in 8th grade science scores, 2009–2015.* www.ecs.org

Engelen, L., Wyver, S., Perry, G., Bundy, A., Chan, T. K. Y., Ragen, J., Bauman, A., & Naughton, G. (2018). Spying on children during a school playground intervention using a novel method for direct observation of activities during outdoor play. *Journal of Adventure Education and Outdoor Learning, 18*(1), 86–95.

Enyedy, N., Danish, J. A., Delacruz, G., & Kumar, M. (2012). Learning physics through play in an augmented reality environment. *International Journal of Computer-Supported Collaborative Learning, 7*(3), 347–377.

Enyedy, N., & Danish, J. (2015). Learning physics through play and embodied reflection in a mixed reality learning environment. In V. R. Lee (Ed.), *Learning technologies and the body: Integration and implementation in formal and informal learning environments.* Routledge.

Fine, S. M. (2014). "A slow revolution": Toward a theory of intellectual playfulness in high school classrooms. *Harvard Educational Review, 84*(1), 1–23.

Finn, J. D. (1989). Withdrawing from school. *Review of Education Research, 59*(2), 117–142.

Fredricks, J. A., Filsecker, M., & Lawson, M. A. (2016). Student engagement, context, and adjustment: Addressing definitional measurement and methodological issues. *Learning and Instruction, 43*, 1–4.

Fredricks, J., McColskey, W., Meli, J., Mordica, J., Montrosse, B., & Money, K. (2011). *Measuring student engagement in upper elementary through high school: A description of 21 instruments* (Issues & Answers Report, REL 2011, No. 098). U.S. Department of Education, Regional Educational Laboratory Southeast. http://ies.ed.gov/ncee/edlabs

Friedman, A. (2003). They're having fun, but are they learning anything? In S. Grinell (Ed.), *A place for learning science: Starting a science center and keeping it running.* Association of Science-Technology Centers Incorporated.

Friedman, L. B., Margolin, J., Swanlund, A., Dhillon, S., & Liu, F. (2017). *Playground Physics impact study*. American Institutes for Research. www.air.org/playgroundphysics.

Galetzka, C. (2017). The story so far: How embodied cognition advances our understanding of meaning making. *Frontiers in Psychology, 8*(1315), 1–5. www.frontiersin.org/articles/10.3389/fpsyg.2017.01315/full

Honey, M., & Hilton, M. (2011). *Learning science through computer games and simulations*. The National Academies Press.

Honey, M., & Kanter, D. (2013). *Design, make, play: Growing the next generation of STEM innovators*. Routledge.

Humphrey, T., Gutwill, J. P., & the Exploratorium APE Team. (2005). *Fostering active prolonged engagement: The art of creating APE exhibits*. Left Coast Press.

Kanter, D. E., Honwad, S., Diones, R., & Fernandez, A. (2013). SciGames: Guided play games that enhance both student engagement and science learning in tandem. In M. Honey & D. E. Kanter (Eds.), *Design, make, play: Growing the next generation of STEM innovators*, (pp. 182–197). Routledge.

Keeley, P. (2016). *Science formative assessment, Vol. 1: 75 practical strategies for linking assessment, instruction, and learning*. Corwin; NSTA Press.

Ladd, W. L., Heral-Brown, S. L., & Kochel, K. P. (2009). Peers and motivation. In K. R. Wentzel & A. Wigfield (Eds.), *Handbook of motivation at school* (pp. 324–348) Routledge.

Lee, C. S., Hayes, K. N., Seitz, J., DiStefano, R., & O'Connor, D. (2016). Understanding motivational structures that differentially predict engagement and achievement in middle school science. *International Journal of Science Education, 38*(2), 192–215.

Lee, V. (Ed.). (2015). *Learning technologies and the body: Integration and implementation in formal and informal learning environments*. Routledge.

Lemke, J., Lecusay, R., Cole, M., & Michalchik, V. (2015). *Documenting and assessing learning in informal and media-rich environments*. MIT Press.

Lindgren, R. (2014). "How do I move this?": The delicate dance of control mechanisms in embodied science learning simulations. In J. L. Polman, E. A. Kyza, D. K. O'Neill, I. Tabak, W. R. Penuel, A. S. Jurow ... L. D'Amico (Eds.), *Learning and becoming in practice: The International Conference of the Learning Sciences 2014* (Vol. 3, pp. 1202–1203). International Society of Learning Sciences.

Lyon, G. H., Jafri, J., & St. Louis, K. (2012). Beyond the pipeline: STEM pathways for youth development. *AfterSchool Matters, Special focus on STEM learning* (pp. 48–57).

Margolin, J., Ba, H., Friedman, L. B., Swanlund, A., Dhillon, S., & Liu, F. (2020). Examining the impact of a play-based middle school physics program. *Journal of Research on Technology in Education*. https://doi.org/10.1080/15391523.2020.1754973

McKenzie, T. L. (2006, January). SOPLAY: System for observing play and leisure activity in youth. *Active Living Research*. http://activelivingresearch.org/node/10642

Mundry, S., & Dunne, K. (2003). *Teachers as learners: Facilitator's guide*. Corwin Press.

National Academy of Sciences. (2011). *Expanding underrepresented minority participation: America's science and technology talent at the crossroads*. The National Academies Press. https://doi.org/10.17226/12984

National Research Council. (2008). *Ready, set, science! Putting research to work in K-8 science classrooms*. The National Academies Press. https://doi.org/10.17226/11882

National Research Council, (2009). *Learning science in informal environments: People, places, and pursuits*. The National Academies Press. https://doi.org/10.17226/12190

Nemirovsky, R., Tierny, C., & Wright, T. (1998). Body motion and graphing. *Cognition and Instruction, 16*(2), 119–172.

Peppler, K. (2017). *The SAGE encyclopedia of out-of-school learning*. SAGE Publications.

Quinn, D. M. & Cooc, N. (2015). Science achievement gaps by gender and race/ethnicity in elementary and middle school: Trends and predictors. *Educational Researcher, 44*(6), 336–346.

Renninger, K. A., & Bachrach, J. E. (2015). Studying triggers for interest and engagement using observational methods. *Educational Psychologist, 50*(1), 58–69.

Robertson, A. D., Atkins, L. J., Levin, D. M., & Richards, J. (2015). What is responsive teaching? In A.D. Robertson, R. Scherr, & D. Hammer (Eds.), *Responsive Teaching in Science and Mathematics* (pp. 1–35) . Routledge.

Rutten, N., Van Joolingen, W. R., & Van der Veen, J. T. (2012). The learning effects of computer simulations in science education. *Computers & Education, 58*(1), 136–153.

Shernoff, D. J., & Vandell, D. L. (2007). Engagement in after-school program activities: Quality of experience from the perspective of participants. *Journal of Youth and Adolescence, 36*(7), 891–903.

Shute, V. J., Ventura, M., & Kim, Y. J. (2013). Assessment and learning of qualitative physics in Newton's Playground. *The Journal of Educational Research, 106*, 423–430.

Singer, D. G., Golinkoff, R. M., & Hirsh-Pasek, K. (Eds.). (2006). *Play = learning: How play motivates and enhances children's cognitive and social-emotional growth*. Oxford University Press.

Skinner, E. A., Kindermann, T. A., Connell, J. P., & Wellborn, J. G. (2009). Engagement and disaffection as organizational constructs in the dynamics of motivational development. In K. R. Wentzel & A. Wigfield (Eds.), *Handbook of motivation at school* (pp. 223–245). Routledge.

Smetana, L. K., & Bell, R. L. (2012). Computer simulations to support science instruction and learning: A critical review of the literature. *International Journal of Science Education, 34*(9), 1337–1370.

Squire, K., Barnett, M., Grant, J. M., & Higginbotham, T. (2004). Electromagnetism supercharged! Learning physics with digital simulating games. In *Proceedings of the 6th International Conference on Learning Sciences* (pp. 513–520). International Society of the Learning Sciences.

Trudeau, J. J., & Dixon, J. A. (2007). Embodiment and abstraction: Actions create relational representations. *Psychonomic Bulletin & Review, 14*(5), 994–1000.

American Institutes for Research. (2016). *STEM 2026: A vision for innovation in STEM education*. U.S. Department of Education.

Voelkl, K. E. (1997). Identification with school. *American Journal of Education, 105*(3), 294–318.

Wang, M. T., & Holcombe, R. (2010). Adolescents' perceptions of classroom environment, school engagement, and academic achievement. *American Educational Research Journal, 47*, 633–662.

Wesiberg, D. S., Hirsh-Pasek, K., & Golinkoff, R. M. (2013). Guided play: Where curricular goals meet a playful pedagogy. *Mind, Brain, and Education, 7*(2), 104–112.

Zohar, R., Bagno, E., Eylon, B., & Abrahamson, D. (2017, July 9–11). *Motor skills, creativity and cognition in learning physics concepts* [Paper presentation]. Movement 2017: Brain, Body, Cognition, First Annual Meeting, Oxford, UK.

chapter 11

Integrating Computational Thinking Across the Elementary Curriculum

A Professional Development Approach

Anthony Negron

At a high-poverty, majority-Latinx public elementary school in Queens, New York, a group of fifth-grade students analyzes a batch of receipts, a toy catalog, and lists of items purchased by certain customers from a toy store. Using this evidence, they try different strategies to figure out which of these customers was the prize-winning "millionth customer." Some strategies prove to be more efficient than others – specifically, strategies that reduce the need for unnecessary calculations by eliminating customers based on the number and costs of the items they purchased compared with the winning receipt. After going through this exercise, the whole class discusses their answers to the problem and the strategies they used. These students are applying a strategy called computational thinking to solve a problem.

In 2018, the New York Hall of Science (NYSCI) undertook a collaboration with teachers at a Queens elementary school to embed computational thinking into their instructional practice. With origins in computer science, computational thinking has evolved into a process for solving problems with broader applications for learning science, technology, engineering, and mathematics (STEM), as well as other subjects. Although computational thinking has various definitions, in our initiative it refers to a way of thinking that includes defining and breaking down a problem into simpler components, finding patterns within the components, removing unnecessary information, and developing step-by-step instructions for solving the original problem and addressing errors. This chapter describes the collaboration, its purpose, the specific actions undertaken in partnership with teachers, and its impact.

The Case for Computational Thinking in STEM Education

Over the past 10 years, both private and public funders have invested heavily in establishing computer science and computational thinking as fundamental components of K–12 STEM education in the United States (Code Advocacy Coalition & Computer Science Teachers Association, 2020; Committee on STEM Education, 2018). The reason for these efforts is obvious. While computing occupations are the number one source of all new wages in the US and make up over half of all projected new jobs, only 47% of US high schools offer computer science courses (Code Advocacy Coalition, 2020). Many of these efforts have supported the creation of curricular and professional development resources that build teachers' and students' understanding of the foundational concepts and practices of computational thinking.

These investments, as well as a range of policy initiatives, have sought to increase students' access to computer science courses or to integrate core disciplinary practices of computer science into existing STEM courses. Various standards for STEM learning, among them the New York Department of Education (2019) blueprint for Computer Science for All (CS4All) and the Next Generation Science Standards, call for the integration of computational thinking concepts and practices into science teaching across the K–12 spectrum (College Board, 2016; National Research Council [NRC], 2013; Villavicenco et al., 2018; Weintrop et al., 2016). These standards documents include broadly defined science and engineering practices, crosscutting concepts, and discipline-specific skills and concepts such as analyzing data, programming, and modeling. They emphasize practices and ways of thinking that are central to computer science and computational thinking and are increasingly driving new discoveries and modes of working in the STEM professions (Grover & Pea, 2013).

While these investments have led to major increases in stand-alone computer science courses (Code Advocacy Coalition, 2020), they have had far less impact on the broader integration of computational thinking concepts and practices into K–12 instruction (Rich et al., 2017). In order to integrate computational thinking into science classrooms, particularly at the elementary level, teachers need high-quality

anthony negron

instructional materials that build appropriate connections between computational thinking and core academic content. They also need professional development that integrates computational thinking practices and concepts into the content they are required to teach (Basu et al., 2016; van Driel et al., 2001; Weintrop et al., 2016). Accomplishing these goals calls for a different approach to curricular design than in freestanding courses and also requires sustained investment in professional development and support.

The NYSCI School Collaboration

NYSCI's collaboration with teachers from a Queens elementary school is part of the Integrating Computational Thinking initiative, which began in 2018 and is supported by the Robin Hood Foundation's Learning + Technology Fund. This initiative is aligned with New York City's citywide CS4All initiative and seeks to address the problem that "no one really knows what integrating computational thinking throughout the K–5 curriculum 'should' look like, or how teachers would implement it" (Robin Hood Learning + Technology Fund, n.d.). The broader Robin Hood initiative connects 21 high-poverty elementary schools in New York City with expert partners to design and implement professional development and computational thinking curriculum in the elementary grades. NYSCI's partner school in the Robin Hood initiative is PS 13Q, a large K–8 school serving 1,300 students. About 75% of the school's students are eligible for free/reduced price lunch, 30% have been identified as English language learners (ELLs), and 56% identify as Latinx. The school is located in Corona, Queens, where NYSCI resides.

Through this partnership, NYSCI has sought to provide early exposure to computational thinking to over 1,000 elementary school students and to empower the 62 educators – the school's entire teaching staff in Grades 2 through 5 – to infuse their work with computational thinking methods. This partnership also aims to improve perceptions of computer science and careers within a community traditionally underrepresented in STEM careers and to broaden learning opportunities for underserved students in a diverse Queens community.

In the following sections, we explain the specific steps we took to build a productive collaboration with the school community and work together on three key aspects:

- Define what computational thinking at the elementary grades can look like when the goal involves integration across the curriculum
- Implement a flexible coaching model with PS 13Q teachers and their leadership team to give educators a central role in developing productive strategies for their students
- Enact an approach to computational thinking that integrates instructional strategies into core content areas to support diverse learners

Codesigning a Definition of Computational Thinking

Computational thinking has been part of computer science for several decades but its applications to K–12 STEM teaching and learning are more recent. Seymour Papert, a founding member of the faculty at MIT's world-renowned Media Lab, is widely credited with being the first person to use the term computational thinking in the 1980s (Papert, 1980). In 2006, Jeannette Wing, a computer scientist who directs Columbia University's Data Science Institute, not only popularized the term in a 2006 paper but also established a foundation for using computational thinking strategies to support learning more broadly in STEM and other domains (Wing, 2006, 2008; Fletcher & Lu, 2009). Computational thinking can help students become active and efficient problem-solvers by drawing on fundamental computer science concepts and practices. It also can provide them with the skills necessary to design simple and elegant solutions to complex systems, such as global environmental change, metabolic functioning, traffic jams, disaster relief networks, and power grid structures.

Further research has illuminated how computational thinking can allow learners, among other things, to formulate and solve problems using digital tools. Specifically, these computational thinking skills include analyzing and representing data through abstractions such as models and simulations; using a series of efficient and effective steps to identify, analyze, and automate possible solutions; and applying this process to solve a wide variety of other problems (Lee, 2015). Since Wing's 2006 paper, scholars have offered many definitions (Grover & Pea, 2013; NRC, 2011; Voogt et al., 2015; Weintrop et al., 2016), each drawing from the researchers' distinctive disciplinary perspectives. More recently, Weintrop and colleagues (2016) conducted a systematic review of computational thinking for mathematics and science classrooms. They identified five broad thinking practices: (a) investigating a complex system as a whole, (b) understanding the relationships within a system, (c) thinking about different levels within a system, (d) communicating information about a system, and (e) defining systems and managing complexity. Our goal was to develop a definition that would be generalizable to domains beyond computer science and applicable to children's learning at the elementary grades.

Although this prior research offered a good starting point, the NYSCI team recognized that to ensure teacher buy-in, we would need to collaborate with our partner educators at PS 13Q to co-construct a definition of computational thinking that they felt would work with their young students. To initiate this project, we built on what we learned through literature reviews to explore existing definitions of computational thinking. NYSCI staff introduced the idea that computational thinking is a problem-solving strategy that is derived from computer science but contains approaches to problem-solving that can apply to any domain. We then defined a set of specific processes that could be used instructionally with our students. Over time, we solidified an operational definition of computational thinking as follows: computational thinking is a problem-solving strategy that is derived from computer science but is also applicable in any domain. This strategy includes the following four core elements:

- *Decomposition.* Breaking a problem into smaller, more manageable parts (such as sub-problems, variables, or categories)
- *Pattern recognition.* Using prior knowledge to find patterns within the smaller problem that will help solve the complex problem more efficiently
- *Abstraction.* Removing unnecessary information and focusing on what is truly important in a given situation

- *Algorithm/debugging.* Developing a series of instructions (an algorithm) to solve the original problem and evaluating the solution to address any errors

Our discussions that produced this definition led to important clarifications. The critical element of debugging, for example, would likely have been overlooked were it not for the experience of and input from the teachers. We realized that debugging is necessary because it addresses situations in which a determined solution is wrong and acknowledges that everyone learns from mistakes and failures (Kumar, 2015). Our collaborators felt this was important for their students to recognize.

Throughout this collaborative work, we maintained a non-negotiable commitment to establishing and sustaining an inclusive culture that valued everyone's insights and perspectives. The process of codesigning a definition with colleagues at PS 13Q was a pivotal first step in this process.

Codesigning a Coaching Model

Teachers have multiple responsibilities each year. Finding time to introduce new curricula, programs, and materials and establish their relevance to classroom practice is always a challenge. Teachers are understandably skeptical about administrative efforts that impose new policies and practices, often without their input. To address these concerns, we emphasized two key points from the outset of the project. First, incorporating computational thinking strategies into existing curricula was not a departure from the skills and concepts that had to be taught on a daily basis. Second, computational thinking strategies can help students become more attentive to the requirements of problem-solving.

To ensure that the work would be relevant to the school's existing curriculum, the NYSCI team conducted extensive research into the educational materials teachers were using. Our goal was to enable teachers to see early on how we could build a bridge between their existing curricula and the computational strategies we had jointly defined. This approach had the benefit of enabling teachers to use lessons they had already planned and materials they had prepared as the core of the computational thinking work. Rather than trying to significantly alter teachers' classroom work, this approach built on their expertise and gave them confidence about integrating new practices.

NYSCI's coaching model in computational thinking draws on years of work by colleagues in the education services division of the museum. This model has been used and refined through work with numerous elementary and middle schools throughout New York City. The model introduces a collaborative process that recognizes teachers' expertise. As described below, the model uses a scaffolded approach that provides greater support to teachers as new instructional practices are introduced and then gradually shifts responsibility for implementation to teachers. It draws on an assets-based approach to professional development that builds new competencies by anchoring them firmly in teachers' existing knowledge.

The coaching process begins with the I-Do technique, in which the NYSCI team models how to integrate computational thinking into an existing lesson. All of the computational thinking lessons used during the I-Do phase are developed with input from teachers and focused on content that teachers had chosen because their students often struggle with it. In PS 13Q, the impact of this first phase was immediate. Teachers experienced firsthand how computational thinking can help students explore challenging content, and students learned strategies they can apply routinely when solving challenging problems. After the implementation of each lesson, time was set aside for reflection. This process often resulted in modifications to activities to help ensure that the lessons worked for the teachers and that students were engaged.

During the second phase of the coaching process, we introduce the We-Do technique, in which teachers take sole responsibility for designing and implementing a computational thinking lesson. The process is similar to phase one, but the lessons incorporate grade-appropriate digital tools that can be used to further support students' understanding of a specific content area. Once again, reflection is built into the process and improvements are implemented.

In the third and final You-Do phase, teachers develop lessons on their own. Once again, they reflect on their experience with colleagues and NYSCI staff and make improvements.

During the PS 13Q coaching, teachers' questions centered on how computational thinking could work across grade levels and how it could be used to accommodate the needs of different kinds of learners. For example, teachers wondered how computational thinking might look different in a second-grade versus a fifth-grade classroom, or how to develop the computational thinking skills of neurodiverse students (those with dyslexia or dyspraxia, or on the autism spectrum) and of ELLs.

Throughout this work we drew upon NYSCI's Design Make Play approach, which calls for engaging learners as creators and makers, helping learners to investigate problems that afford multiple ways for solving them, and developing inviting activities with a variety of entry levels. (See this book's Introduction for more details about the Design Make Play principles.)

The coaching model, combined with NYSCI's Design Make Play approach to learning and engagement, ensures through iterative cycles of reflection and improvement that all stakeholders play a role in shaping lessons that meet desired outcomes.

Integrating Computational Thinking into Student Activities

To build students' foundational understanding of how to use computational thinking to solve instructional problems, we introduced the idea of design challenges. As part of this process, we used reflective conversations to help students see connections between their daily lives and their schoolwork. We codeveloped activities with their teachers and iterated on them through multiple implementation cycles. Below, we illustrate some ways to design activities that have relevance to students' lives.

First, we found that using a task like cleaning a messy room created a context that all students could identify with. Providing everyday examples of computational thinking establishes a personal connection that helps students better understand why they are using these strategies in their schoolwork.

Second, to best accommodate the needs of diverse learners, we introduced each computational thinking component separately. This gives students the opportunity to practice one skill set at a time.

Third, for early learners (K–2), we found it especially beneficial to relate the components of computational thinking to their regular activities at home, in school, and during play. This helps children form positive associations and gain familiarity with computational thinking components. For example, we introduced "algorithms" as a vocabulary term when students were learning to wash their hands after an art activity. We then had them create a series of steps for hand washing, which was a powerful way to anchor an abstract term in familiar practices.

It is also helpful to look for opportunities to practice foundational computational thinking skills within the structure of a normal classroom day or within specific lessons that target early childhood skills such as sequencing, sorting, and pattern identification. One lesson designed for a second-grade class, for example, focused solely on pattern recognition, which tied directly to the standard of supporting students' ability to understand informational text. For that lesson, we first asked students to say what would come next in a sequence of images and identify patterns in the classroom. We then read a portion of the book *Friends Around the World* to them and stopped after the first few pages. We asked students to identify patterns in the sections of the story read to them, predict what would happen next, and provide evidence to support their predictions. Students were able to correctly identify how the story continued and where they could look for text features in the book to support their understanding.

Fourth, displaying charts or posters with computational thinking vocabulary throughout the classroom or school helped students to keep key concepts in mind and gave teachers something concrete to point to. By developing their own posters and charts, teachers could connect computational thinking vocabulary to specific classroom activities and daily tasks.

Fifth, incorporating questions about computational thinking strategies (see Table 11.1) into daily classroom facilitation across all subject areas served to reinforce a

Table 11.1 Computational thinking questioning strategies

Computational thinking core elements	Computational thinking questions
Decomposition	How can you break this problem down into smaller steps?
Pattern recognition	What patterns do you notice that may be helpful in solving this problem? What do you already know about this subject that will help?
Abstraction	What are the important and unimportant parts to know to solve the problem?
Algorithm/debugging	What are the specific steps that you need to take to solve this problem? How do you test it out to make sure these steps are correct and fix it if needed?

computational thinking mindset among students. When teachers used the questions described in Table 11.1, it not only helped reinforce specific elements of computational thinking but also helped teachers and students take a more intentional approach to problem-solving across the curriculum.

The Case of the Missing Winner

To further illustrate how we implemented this approach to computational thinking in elementary classrooms, we use the example of an activity that was codeveloped with fifth-grade math teachers who wanted to focus on adding and subtracting decimals. With either of these functions, the biggest challenge for their students was making sure to align the decimals when performing either of these functions.

In the first part of the lesson, students were given a worksheet with an algorithm for adding and subtracting decimals, but with the steps out of order. To solve this problem, students needed to take at least three steps. First, if they were adding and got a two-digit answer in one column, they needed to "carry the 1" into the next column on the left. If they were subtracting and the digit being subtracted in a column was larger than the digit above it, they had to "borrow" from the next column on the left. Second, they had to vertically stack the numbers they were working with. Third, they had to align the decimal points. Students were then asked to reorder the steps in a sequence that they thought made sense. Once they rearranged the steps, they engaged in a group conversation to share the steps (algorithms) each student came up with and make any necessary modifications to their algorithm. In this way, students were able to identify which sequence worked best for them and to learn from their peers that there are alternative ways to solve a problem.

In the second part of the lesson (Figure 11.1), students tested their algorithm with a fun narrative called The Case of the Missing Winner. In this narrative, students had to figure out who was the mysterious millionth customer while using only the evidence provided to them. The evidence included a receipt, a list of items purchased by each potential winner, and a toy catalog with prices of all the toys that could be bought at Dylan's Toy Shop. NYSCI and PS 13Q staff listened to the different ways in which students drew connections to computational thinking to solve the problem and applied it during the lesson.

Some students totaled the costs of all the items purchased by each of the potential winning candidates. Although this was not the most efficient way of approaching the problem, it still allowed students to answer the question. Other students used the information from the winning receipt to eliminate two of the potential winning candidates, who had too many or too few items that did not match the winning receipt. After this elimination process, students only had to total the items of the three remaining candidates to arrive at the correct answer. With this latter approach, students were applying the computational thinking strategy of abstraction – removing unnecessary information to reach the correct answer more efficiently.

A final approach used by some students was similar but included an additional strategy. While these students were adding and reviewing individual item prices, they noticed that some prices exceeded the total amount of the winning receipt, so they could eliminate potential winners who had bought these more expensive items. This

Figure 11.1 PS 13Q fifth-grade students working on The Case of the Missing Winner activity under the supervision of NYSCI's manager of digital programming.

was the most efficient way of tackling the missing winner challenge, and it helped to build students' abstraction skills. When students arrived at their final answers, the whole class had a conversation about who the mysterious person was and how the students used computational thinking to solve this problem.

Charlotte's Web Computational Thinking Activity

One of the many concerns raised by teachers was whether computational thinking would be a good fit for students with varied learning needs. This was a particular concern among teachers of ELLs, neurodiverse students, and children under eight years of age. To illustrate how this challenge can be addressed, we use the example of an activity designed during the You-Do portion of the coaching model by one of the second-grade general education teachers at PS 13Q, Mrs. Lee (a pseudonym). At the outset, she had reservations about incorporating computational thinking based on the age and diversity of learning needs among her students.

Mrs. Lee focused her attention on integrating computational thinking into an English language arts lesson with the following learning objective: How can readers identify the main events of a story in sequence? During her design process, she made sure to brainstorm with her peers and involve them in refining the lesson. She also familiarized herself with the ScratchJr digital tool that her students would use for the lesson. At the start of the lesson, students had to use their computational thinking skills to sequence the events within two specific chapters of the book *Charlotte's Web*. Students were grouped based on the levels of support the teacher felt they needed:

- *Group 1 (support)*. Students are homogeneously grouped and receive additional teacher support to complete the task using sequence cards with pictures.

- *Group 2 (on level)*. Students are heterogeneously grouped to complete the task as instructed.
- *Group 3 (enrich)*. Students are heterogeneously grouped to complete the task using more detailed sequence sentences.

The grouping was intended to ensure that students with different needs received an equitable experience so they could successfully achieve the learning goals and objectives. After working in small groups, students engaged in whole-class discussion to share which events they had identified as "main" and how these events could be correctly sequenced in the narrative. After this discussion, students were asked to use ScratchJr to digitally recreate the key parts within the main events they had just put in order. This process helped students to remove unnecessary information and focus on the key details of the story.

For example, as part of this activity, students had to focus on a chapter with a time span of several years. Most students programmed key moments in the chapter, including when the character Wilbur returns home to the barn and when Charlotte's tiny spiders are born. We observed students correctly sequencing the main events from the book and drawing from certain text features like images and headings to support their ideas. This project had the added benefit of introducing them to new algorithms in the form of the coding blocks needed to create projects in ScratchJr. Overall, Mrs. Lee's lesson was a prime example of how computational thinking could meaningfully support students' learning in subjects other than STEM and how this project was truly starting to change how teachers approached their lesson designs and class facilitation.

Overall Impact of Project

At the start of this project, NYSCI staff, in collaboration with our PS 13Q partners, developed a working definition of computational thinking. We anchored this work in previous definitions but adapted it to be appropriate for elementary age children and instructionally relevant for their teachers. The intention was to shape accessible and flexible strategies for both teachers and students that could be applied across subject areas.

As the project unfolded over a two-year period, it was encouraging to see teachers integrating computational thinking practices into their classrooms for all types of learners. Significantly, the work we did together led teachers and students to use computational thinking as a broader problem-solving approach.

Toward the end of the project, we reviewed all of the teachers' reflection videos and surveys for evidence of impact. PS 13Q teachers said that the coaching model helped them to develop agency and feel competent and confident in using computational strategies throughout their practice. The teachers also reported that the more they embedded computational thinking into their routines, the more they noticed their students taking ownership of their own learning, improving their understanding of the content of and connections between different subjects, and developing a belief in their own abilities to tackle complex problems.

anthony negron

The Education Development Center Inc., the research partner for this project, studied our professional development work with PS 13Q. They also did case studies of other professional development models that supported computational thinking integration efforts in more than 20 different schools. Preliminary findings from this work show that a consistent challenge for integrating computational thinking across multiple schools and grade levels is the lack of instructional support materials, particularly strategies for identifying emergent computational thinking that are appropriate for elementary students (Yan et al., 2020). When asked to define computational thinking, teachers routinely characterize it as a metacognitive strategy that allows for in-depth learning and helps students think systematically to solve problems. The challenge, however, is that it is difficult for teachers to use typical assessment measures, such as tests or quizzes, to gather evidence of students' computational thinking and its application to content in core subjects. While teachers reported examples of their students' drawing on computational thinking concepts and practices during core subject work, teachers did not have replicable, consistent techniques to identify how students used distinct concepts and practices as a part of their problem-solving. We plan to address this shared challenge during the next phases of our work.

Conclusion

The work summarized in this chapter demonstrates the potential of working collaboratively with teachers to codesign strategies for implementing computational thinking skills at the elementary level. We intend to expand this work to help many more teachers of young children identify and support their students in using computational thinking practices in their classrooms. To address their challenges, NYSCI, in collaboration with our other partners, will codevelop and pilot an online community of practice. This community will provide early elementary educators with opportunities to dialogue with peers and experts and engage in sustained analysis of students' computational thinking and evidence of their learning. Because these conversations will be supported through an online community of practice, they will be documented and transparent. These documented conversations will generate an archive of evidence for future participants to build on and a sustainable approach to supporting communities of practice (Wenger, 2010). This research and development work will set the stage for scaled-up efforts and further study focused on building communities of practice across schools and potentially across districts. This type of reflective inquiry and ongoing examination will help teachers deepen computational thinking instruction with elementary students and integrate it across the curriculum.

References

Basu, S., Biswas, G., Sengupta, P., Dickes, A., Kinnebrew, J. S., & Clark, D. (2016). Identifying middle school students' challenges in computational thinking-based science learning. *Research and Practice in Technology Enhanced Learning, 11*(13), 1–35.

Code Advocacy Coalition & Computer Science Teachers Association. (2020). *2020 state of computer science education: Illuminating disparities.* https://advocacy.code. org/

College Board. (2016). *AP computer science principles: Including the curriculum framework.* https://apcentral.collegeboard.org/courses/ap-computer-science-principles

Committee on STEM Education. (2018). *Charting a course for success: America's strategy for STEM education.* Executive Office of the President of the United States, National Science & Technology Council. www.whitehouse.gov/wp-content/uploads/2018/ 12/STEM-Education-Strategic-Plan-2018.pdf

Fletcher, G. H. L., & Lu, J. J. (2009). Education: Human computing skills: Rethinking the K–12 experience. *Communications of the ACM, 52*(2), 260–264.

Grover, S., & Pea, R. (2013). Computational thinking in K–12: A review of the state of the field. *Educational Researcher, 42*(1), 38–43.

Kumar, M. (2015). Learning from productive failure. *Learning Research and Practice, 1*(1), 51–61.

Lee, I. (2015, February 4). CSTA computational thinking task force. *The Advocate.* Computer Science Teachers Association. http://advocate.csteachers.org/2015/02/ 04/csta-computational-thinking-ct-task-force/

National Research Council. (2011). *Report of a workshop of pedagogical aspects of computational thinking.* Committee for the Workshops on Computational Thinking. The National Academies Press. www.nap.edu/catalog/13170/report-of-a-workshop-on-the-pedagogical-aspects-of-computational-thinking

National Research Council. (2013). *Next Generation Science Standards: For states, by states.* The National Academies Press. www.nap.edu/catalog/18290/next-generation-science-standards-for-states-by-states

New York City Department of Education. (2019). *Computer science for all.* https://blue-print.cs4all.nyc/

Papert, S. (1980). *Mindstorms: Children, computers, and powerful ideas.* Basic Books.

Rich, K. M., Strickland, C., Binkowski, T. A., Moran, C., & Franklin, D. (2017). K–8 learning trajectories derived from research literature: Sequence, repetition, conditionals. *ACM Inroads, 9*(1), 182–190.

Robin Hood Learning + Technology Fund. (n.d.). *Computational thinking concept paper.* www.robinhood.org/uploads/2018/08/Computational_Thinking_Concept_ Paper_2018.pdf

van Driel, J. H., Beijaard, D., & Verloop, N. (2001). Professional development and reform in science education: The role of teachers' practical knowledge. *The Journal of Research in Science Teaching, 38*(2), 137–158.

Villavicenco, A., Fancsali, C., Martin, W., Mark, J., & Cole, R. (2018). *Computer science in New York City: An early look at teacher training opportunities and the landscape of computer science implementation in schools.* Research Alliance for New York City Schools.

Voogt, J., Fisser, P., Good, J., Mishra, P., & Yadav, A. (2015). Computational thinking in compulsory education: Towards an agenda for research and practice. *Education and Information Technologies, 20*(4), 715–728.

Weintrop, D., Beheshti, E., Horn, M., Orton, K., Jona, K., Trouille, L., & Wilensky, U. (2016). Defining computational thinking for mathematics and science classrooms. *Journal of Science Education Technology, 25*(1), 127–147.

Wenger, E. (2010). Communities of practice and social learning systems: The career of a concept. In C. Blackmore (Ed.). *Social learning systems and communities of practice* (pp. 179–198). Springer-Verlag.

Wing, J. (2008). Computational thinking and thinking about computing. *Philosophical Transactions of the Royal Society, 366*(1881), 3717–3725.

Wing, J. M. (2006). Computational thinking. *Communications of the ACM, 49*(3), 33–35.

Yan, W., Liu, R., Israel, M., Sherwood, H., Fancsali, C., & Pierce, M. (2020). School-wide integration of computational thinking into elementary schools: A cross-case study. In J. Zhang & M. Sherriff (Chairs), *Proceedings of the 51st ACM Technical Symposium on Computer Science Education (SIGCSE '20)* (pp. 1325–1325). Association for Computing Machinery. https://doi.org/10.1145/3328778.3372623

The Pack
Playfully Embodying Computational and Systems Thinking

*Leilah Lyons, Stephen Uzzo,
Harouna Ba, and Wren Thompson*

In The Pack, an immersive computer video game developed and tested by the New York Hall of Science (NYSCI) and its partners, children use an avatar, along with virtual "creatures" that can dig, build, and do other functions, to explore the effects of change on virtual ecosystems. In the version of this game that emerged from several iterations of brainstorming, design, testing, and proto-typing, players discover new creatures and entice them to join their "pack." They can "stack" these creatures sequentially into a sort of totem pole to create a set of ordered instructions – an algorithm – that they can use to transform the simulated world and solve complex environmental systems challenges.

These children are doing more than playing. The game was designed to engage children from underserved groups in learning the habits of mind and problem-solving skills that characterize computational thinking (CT) and are essential in the practice of science, technology, engineering, and mathematics (STEM).

To thrive in a world where digital technologies and automated computation are transforming nearly all aspects of human society, all students need the kinds of computational thinking skills and habits of mind that allow them to innovate and solve

leilah lyons et al.

problems more effectively and efficiently. Although frameworks and tools to integrate CT into teaching and learning have advanced significantly over the past 40 years, there is a rapidly widening gap in equitable access to this kind of preparation (Scott et al., 2016). Students from economically disadvantaged and underserved groups, as well as teachers and students in low-income school districts, lack access to computational resources, and gender and racial disparities persist (Grover & Pea, 2013; Repenning et al., 2010; Salac & Franklin, 2019; Scott et al., 2016; Torres-Torres et al., 2020). Merely offering more opportunities to engage in CT may not close these gaps. There is growing evidence that students from demographics already underrepresented in computing fields continue to self-select out of CT learning opportunities (Pinkard et al., 2017).

At NYSCI, we argue that another approach is needed to invite underrepresented students to build CT competencies. Our proposed approach positions CT as a tool that allows learners to achieve other goals, such as solving STEM-related problems. To make the prospect of solving STEM problems with CT appealing to learners, we have embedded STEM and CT concepts within an immersive, playful learning game. Learners become actors with agency in a simulated game world (see Figure 12.1). They learn about STEM and CT concepts through playful trial and error, in which their mistakes reveal as much about the underlying concepts as their successes.

To investigate this approach, NYSCI designed and tested The Pack, a single-player, open-world video computer game that aims to support the agency of all learners through play and embodiment. As used in this chapter, "embodiment" refers to the extension of the learner's thinking into the virtual game world, where tasks are accomplished by interacting with simulated objects or entities that themselves embody computational or complex systems concepts. In this game, learners practice building algorithms associated with CT to solve complex challenges related to

Figure 12.1 Screenshot from the second prototype, depicting the player's avatar standing in front of an algorithm.

ecological systems. By incorporating CT into a STEM learning context, we hope to engage learners who are underrepresented in STEM while addressing national educational practices that call for embedding computational thinking in STEM (dubbed CT-STEM) (Weintrop et al., 2016).

This chapter describes the iterative design and research processes we used to explore how to meaningfully integrate CT in a STEM learning context. We also discuss how prior research informed our work, how learners responded to two iterations of the game, and how our design and research processes revealed nuances about the differences between CT-STEM and more traditional conceptions of CT.

Context and Background

The Pack is an outgrowth of a computer-based museum exhibit called Connected Worlds that immerses visitors in an interactive simulated ecosystem. Visitors can manage the system and its four distinct biomes through gestures. As they interact with the exhibit, they can virtually experience the consequences of their decisions on ecological phenomena and explore concepts that underlie complex systems. (Chapter 4 describes Connected Worlds in more detail.)

The Pack represents an effort to combine the design principles of the Connected Worlds exhibit with CT-STEM concepts in a standalone computer game that could be used to broaden access to computing for students ages 9–16. As our team of informal science educators, game developers, science content experts, and researchers worked together to design and iteratively test The Pack, we uncovered a key design challenge. Visitors in Connected Worlds are motivated to engage with the exhibit because of their desire to explore the simulated ecosystem, but how can we translate this desire into engagement with CT? In other worlds, how can we make computational thinking integral to their exploration without it becoming a barrier to entry?

Computational Thinking Definitions and Skills

In developing The Pack, we drew from decades of research about computational thinking. Seymour Papert (1980) and his colleagues characterized computational habits of mind and reasoning as extensions of thinking and, as a corollary, the manipulation of computing machines as a kind of extended or embodied cognition. This led to their watershed work in developing the LOGO programming language for children. They coined the term "computational thinking" as a way to extend human cognition into machines by reasoning about how machines respond to and represent human interactions.

Expanding on this work, Jeannette Wing (2006) defined CT as "thought processes involved in formulating problems and their solutions so that the solutions are represented in a form that can be effectively carried out by an information-processing agent" (Cuny et al., 2010, cited in Wing, 2011, p. 20). Selby and Woolard (2014) defined computational thinking as the "focused approach to problem-solving" (p. 5) used by computer scientists. This approach incorporates abstraction (removing unnecessary information and focusing on what's important), decomposition (breaking a problem into smaller, more manageable parts), algorithmic design (developing step-by-step instructions), evaluation (evaluating the efficiency and outcomes of processes), and generalization (expanding from a specific to a broader applicability). (Chapter 11 includes a more detailed discussion of the definitions and thinking processes associated with CT.)

Engaging learners in CT is another matter. There seems to be general agreement that CT learning environments (often a programming environment) should have a "low floor, high ceiling" – meaning that they are accessible to novices but continue to be useful as learners master skills. Attaining this goal with traditional programming environments can be challenging (Grover & Pea, 2013). Visuospatial approaches, which emphasize visual perceptions and manipulation of the spatial relationships among objects, have become a common and proven technique to help overcome obstacles to computer programming and aid users in thinking about problems through a computational lens (Kynigos & Latsi, 2007). These approaches have sought to address more complex kinds of computational problems and support teaching and learning about complex modeling systems (Resnick, 1994; Tisue & Wilensky, 2004). Visuospatial approaches are also important to learning with graphical authoring tools (programs used to write hypertext or multimedia applications) such as Scratch (Resnick & Silverman, 2005). These authoring tools use procedures common to programming languages but have human-computer interfaces that emphasize the processes for solving problems computationally. Scratch has enabled children aged 8–11 to successfully use core CT practices like if-then statements (Maloney et al., 2008). However, learners still need to be given reasons to engage with these visuospatial programming tools for understanding STEM concepts.

CT-STEM for Engagement

As science has become increasingly complex and data-driven, the need for computational thinking is more crucial than ever (Weintrop et al., 2016). This has prompted calls to recognize the intersection between CT and STEM disciplines, which embraces skills and habits of mind that go beyond those seen in traditional CT:

- *Data and information.* Collecting, creating, manipulating, analyzing, and visualizing data
- *Modeling and simulations.* Using computational models to understand a concept, understanding how and why computational models work, assessing computational models, using computational models to find and test solutions, and building new computational models or extending existing ones
- *Computational problem-solving.* Troubleshooting and debugging, programming, choosing effective computational tools, assessing different approaches and solutions to a problem, developing modular computational solutions, using problem-solving strategies, and creating abstractions
- *Systems thinking.* Investigating a system as a whole; understanding the relationships within a system; thinking in terms of the different levels that arise from interactions of the parts of a system; visualizing systems; and identifying, understanding, and managing complexity

CT-STEM has the potential to engage learners who might not otherwise find computing interesting; these learners may engage in computing in order to understand a STEM topic rather than for its own sake. For example, research shows that CT educational experiences that encourage problem-solving as part of non-computational challenges may promote self-efficacy, particularly for girls (Ardito et al., 2020). A number of programs aimed at reaching underserved learners have embedded computing in activities as diverse as crafting (Kafai et al., 2014), writing (Burke & Kafai, 2012), and music (Bell & Bell, 2018).

Much like these CT activities in the arts, CT-STEM positions computation as a *tool* that helps learners complete other tasks embedded in authentic STEM practice. Importantly, this shift from CT as a subject to be learned to a tool that can be configured to learn other content suggests that the process for designing a CT-STEM game should not be based around core CT practices (like pattern recognition, abstraction, decomposition, or debugging) or subskills (like sequencing, conditionality, or loops). Rather, the game should be designed to create a learning context in which the STEM content would drive decisions about which computational practices to include.

From its inception, The Pack was intended to emphasize the intersection of CT with reasoning and learning about complex environmental systems. In designing The Pack, we sought to create a context in which learners would find CT useful for exploring the STEM content of complex environmental systems. We also sought to support the agency of all learners. Traditional CT often involves transferring the learner's agency to a computational agent; learners clearly communicate their intent and then allow the computational agent to execute the actions. By contrast, CT-STEM is about using computational tools to solve STEM problems. CT-STEM places more continuous agency in the hands of learners; they are empowered to decide when and where they will use the CT tools.

CT and Complex Systems Thinking

The development of computational thinking skills is key to understanding complex systems in environmental science, such as global climate change and the ecology of ecosystems. Complex systems typically exhibit four properties (Guckenheimer & Ottino, 2008):

- An internal structure consisting of many interacting components and subsystems
- Emergent behaviors that are not predictable or observed in "simple" systems
- Adaptation to inputs and feedbacks over time
- Uncertainty in predicting and controlling aspects of the system or the system as a whole

To illustrate these principles, several computational tools have been developed, usually in the form of simulated microworlds that allow learners to manipulate the rules governing the behavior of agents in the system (Yoon et al., 2018). Research has shown that even elementary-age children can construct complex system simulations with modeling tools such as StarLogo (Klopfer et al., 2005). Studies have also found that

leilah lyons et al.

working with complex systems simulations such as NetLogo can change how learners reason and incorporate CT practices (Resnick & Wilensky, 1998). The challenge with having children construct and interact with complex system simulations in classrooms is that even classroom teachers need extensive professional development to use these tools (Yoon et al., 2017). Prior research has shown the promise of using embodied interfaces, in which learners interact with technology in physically natural ways (i.e., through gestures, for simulations of complex environmental systems) (Shelley et al., 2011). With our project, we set out to explore whether we could produce an embodied educational experience that would have a low barrier to entry for novices but would meaningfully engage middle-school-age learners in CT-STEM practices.

Digital games have been created as a way for learners to engage in systems thinking in the context of environmental science (National Research Council, 2011). Games can embed learners in an embodied way within an immersive environment that gives them agency over their explorations. The EcoMuve and EcoMobile curriculum, for example, situates learners in an immersive 3D simulation of an ecosystem or in a real-world ecosystem via augmented reality to scaffold their development of systems thinking and environmental literacy (Grotzer, 2012). EcoMuve aims to draw out students' causal assumptions about how different parts of the world work and which conditions can cause change, and to see how particular causal dynamics play out in the ecosystem "surrounding" them. While EcoMuve offers virtual world situations based on real experiences in the physical and natural world, it positions the learners primarily as passive documenters and observers, not as *actors* in the system. Although this may work well as part of a school curriculum, it is not inherently engaging to learners – it doesn't feel like a game in which players can affect, and have their decisions affected by, the simulated world.

In designing our game, we wanted to maximize players' engagement with real-world concepts as they played the game. We wanted to construct an experience in which learners would act in an embodied fashion within the game environment and be able to see the impact of their actions on the simulated ecosystem. Further, we wanted the need for players to manipulate the simulated ecosystem to motivate them to use CT. We saw a particular opportunity for learners to use CT in their explorations of complex causality, whereby learners engage in looking for mechanisms that explain why two or more entities are related (Grotzer, 2012).

Design Process

We wanted computational thinking to be a natural and necessary outgrowth of learners engaging with a real challenge in a complex ecological system, and this had implications for our design process. We needed a process that would allow for the reciprocal evolution of our ideas about the ecological content, its potential intersection with CT, and the game mechanisms that would help system properties emerge and support CT. Thus, we opted to use an iterative design process that brought together STEM experts and interactive experience designers. This group met regularly to collectively propose, explore, prototype, critique, and test game ideas.

Design Partners

Our specialists in the scientific domains were members of the Center for International Earth Science Information Network (CIESIN), a research unit of the Earth Institute at Columbia University. CIESIN is a world-class data and analysis center that focuses on geospatial research and practical applications related to the human dimensions of environmental change. Their role in the design process was to review and brainstorm about how complex ecosystems respond to disturbances in equilibrium over time, how interventions would affect those ecosystems, and how these effects would manifest themselves.

Another partner was Design I/O, a creative studio specializing in the design and development of cutting-edge, immersive, interactive digital installations, open world games, and storytelling for events and cultural institutions. As creators of the Connected Worlds exhibit, they observed years of visitor engagement and had a sense of which design factors in that exhibit were most critical to preserve and translate to The Pack learning environment. Their role in The Pack was to iteratively develop multiple designs and prototypes and be deeply involved in all stages of the design and testing environment.

Six middle school teachers, whom we referred to as Design Fellows, and approximately 100 additional teachers were significant collaborators in the formative research and development phase of The Pack. They reviewed the game prototypes and provided feedback to the technology design and science content teams. Their role was to provide insight into teachers' perceptions of computational and systems thinking, their understandings of how the game supports CT, and their interpretations of the usefulness of the game design in addressing teaching challenges or learners' needs. They were also called upon to develop a set of cases showing how The Pack could be used to frame children's investigations and teach relevant concepts or skills, and to foresee possible challenges.

Iterative Development Process

NYSCI researchers conducted weekly testing and reported results to Design I/O team members, who would rapidly respond to the insights gained from the testing and provide updated designs to test as the game evolved. We held a combination of monthly brainstorming meetings and design reviews that included different partners, depending on the topic. We would quickly summarize feedback from these meetings and bring it to the whole team.

The six teacher Design Fellows were involved in many aspects of the design process. They participated in meetings to develop the initial concept, followed by bimonthly or quarterly game critique sessions, focus groups on curricular connections, scenario-building, individual clinical interviews, and four online brainstorming sessions punctuated by small scale trials of using the game with their students. We also conducted five focus groups with 100 teachers (about 20 teachers per focus group) to validate and generalize the findings from the Design Fellows team. We prompted these focus group participants to share their thoughts about CT and environmental systems to get a sense of how their perceptions compared with those reported in the research literature.

Overall, the participants identified specific aspects of CT, such as decomposition, looking for patterns, and sequencing. They also understood that CT involved more general processes and practices such as problem-solving, collaboration, trial and error thinking, math, and data. For environmental systems, their examples of topics were very connected to ecological concepts included in the Next Generation Science Standards for middle school and some high school science courses. Their biggest challenge lay with identifying intersections between CT and environmental systems. They generally noted that all of the computational skills they identified could be used to address environmental concepts or ideas but did not elaborate on how this could be implemented or where those connections were.

Over the course of the 24-month design and development process, researchers from NYSCI worked together with all of the partners to produce and test over a dozen potential game ideas. Two of these ideas were developed into functional prototypes that we could test with middle school students. Each of the prototype studies explored different questions about CT, CT-STEM, and player engagement, reflecting what we had learned. The first prototype allowed us to explore how game mechanics might help players engage with complex causality and whether players would use CT to explore the causality present in the simulated system. With the second prototype we explored how to support players' engagement with the game's computational thinking features and whether integrating computational elements into the simulated system would support players' understanding of both computational and complex systems principles.

Prototype-Testing with Players

We tested the two game prototypes with 32 participants, aged 9–16. While our institutional review board did not permit us to collect individual demographic information, our participants were drawn from the general population of our museum, which in 2019 was 32% White, 19% Hispanic/Latinx, 18% Asian/Indian, 10% two or more races, and 9% Black/African American. We conducted informal observations and used QuickTime screen capture to record the laptop screen and associated conversations as players interacted with each prototype and verbalized their thinking. In addition, we interviewed 25 individual participants. The interviewer also took notes about the approaches taken as the player progressed through the game.

We made videos of game play and had transcriptions made of the audio from these videos. We then reviewed the transcripts while watching the videos and added details to the transcripts when necessary to note what was happening on the game screen at the time of a conversation. We also added details from our observation notes when necessary. We analyzed the data to highlight players' engagement with the game environment and determine how players' engagement supported their understanding of computational and complex systems principles.

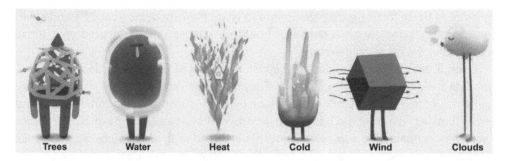

Figure 12.2 Depictions of the six different elemental characters.

Design of Iteration 1

The first iteration was structured to give learners the opportunity to explore multiple open-ended environmental outcomes without asking them to program the behaviors of agents in the game. The designers produced an interactive ecology into which different "elemental" avatars – representing trees, water, heat, cold, wind, and clouds – could be placed (see Figure 12.2). Three of each type of avatar were available to add to the environment. Each avatar affected the environment in different ways – by adding plant seeds, adding moisture, adding or "removing" heat energy, or adding atmospheric velocity or evaporation (see Figure 12.3). Some of the elements function in relation to each other; for example, fire and ice cancel each other out, and wind blows clouds and seeds. The number of avatars of the same type of element that are added (from zero to three) has an effect; adding more avatars multiplies the impact of an element, although some combinations are more noticeable than others. Using the maximum number of water elements will result in a flood that destroys vegetation, if not countered by cloud elements, which cause the surface water to evaporate. Maximum wind, on the other hand, makes the trees and grasses move more dramatically by dispersing seeds but has little effect on the climate.

In this iteration, the complex systems concepts were embodied in the game, while the computational thinking was left to the learners. To explore the space of possible environmental outcomes more effectively, learners would need to organize their tests in a systematic fashion. Learners have been documented responding to such an open-ended challenge by inventing a legitimate algorithmic approach – for example, by using a strategy of controlling variables (Moher, 2009) – although we suspected that the CT aspect would need to be highly facilitated.

Player Responses to Iteration 1

Our first study investigated whether Iteration 1's mechanics might help players engage with complex causality, and whether players would use CT to explore the causality present in the simulated system. Complex causality differs from simple causality, in which an event is either a cause or an effect. The more complex form of causality featured in Iteration 1 is chain causality, in which an event is both the effect of the previous cause and the cause of the subsequent effect (Grotzer, 2012). Some researchers

leilah lyons et al.

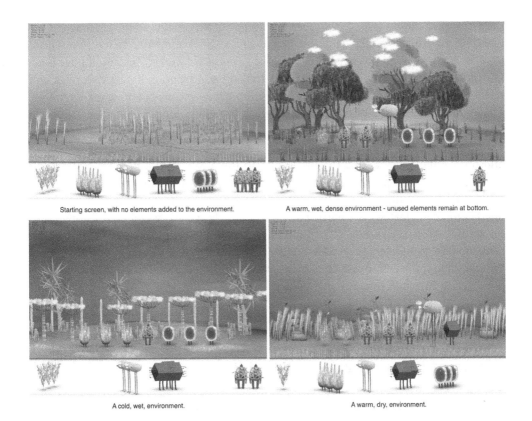

Starting screen, with no elements added to the environment.

A warm, wet, dense environment - unused elements remain at bottom.

A cold, wet, environment.

A warm, dry, environment.

Figure 12.3 Screenshots of the first prototype, showing how the environment changes as different numbers of different elemental creatures are added to it.

have hypothesized that reasoning about chain causality may serve as a bridge for novice learners, helping them move beyond a surface-level ability to recall a sequence of events to using "domino" logic, in which they understand how a prior effect made a subsequent event possible (Grotzer, 2012). Our study examined what kind of causal logic children would use in playing with Iteration 1 of The Pack and if the nature of the causality in the game encouraged them to use CT.

Regardless of age, the most common approach to the game consisted of testing each element one by one without adding any other elements. In testing the game, younger children (11 and under) showed that for them, causality was direct and simple. If they acted, there was an immediate and visible consequence – or there was none. Their process of identifying the effect of each element was to place it on the screen one at a time. When they had placed one element and saw little change, they added another element to see what it would do. It did not occur to them that they were now seeing the effect of two elements rather than of just the last one they added. *They did not think about the state of the system as the result of a chain of events, as in domino logic, or as the result of an interaction of multiple events.* The children knew that the wind blows the clouds around, for instance, but they did not relate it to the idea that the clouds produce rain.

By contrast, early adolescents (ages 12–13) in our target age range approached the task as if it were a natural part of learning any new game; they tried to figure out *what they could do in the game rather than how the game world works*. This could have been an opportunity for CT to emerge, in that making connections between causes and effects is a problem amenable to decomposition. However, we did not see CT emerge. These adolescent learners played around with the game for a while to notice which actions made other things happen. When they had a hunch, they tried it a few times but paid little systematic attention to the relationship between the state of the world and the element they added or subtracted. In the absence of a goal beyond exploration, several participants invented stories to the effect that the object of the game was to develop a superpower by combining elements, so that an (imagined) antagonist could be defeated. The participants were clearly seeking a reason to engage with the game. While their in-game actions affected the state of the environment, these actions were completely reversible and free of consequences. The game put them in the role of an observer, not a consequential actor. They wanted to be able to accomplish a task that had more challenge and more reward than the pure exploration of system causality. This made us realize that a fundamental redesign of the game was needed.

Design of Iteration 2

The fact that children in our target age range approached Iteration 1 by focusing primarily on what they could do in the game, rather than on how the game world worked, suggested that for Iteration 2 we needed to place more emphasis on learner actions and their consequences. Thus, for Iteration 2, we wanted to shift the design from embodying aspects of the complex environmental system to embodying the process of computation, since that offers more possibilities for direct player manipulations.

The interviews from Iteration 1 also revealed that, even though participants enjoyed the aesthetically pleasing simulated environment, they did not feel immersed in it in any meaningful way. Little about the context inspired them to explore it deeply. This suggested the need for additional narrative framing, much like the narrative framing that engages Alice users (a computer programming environment developed by Kelleher et al., 2007). If exploring a complex system is the intended goal of the game, we had to construct a story for why the player would need to pursue this goal.

Iteration 2 placed a light narrative on top of the core exploratory game to engage learners more deeply in the problem context. It is an interactive, open-ended digital game in which players explore a dynamically generated set of ecosystems – in other words, open worlds that change based on the player's behavior in the game – and face incrementally more complex challenges. During their journeys, they discover new creatures and entice them to join their "pack." Each creature performs a specific function that can be combined sequentially with other creatures to form reusable algorithms that can transform the natural world around them; for example, by digging

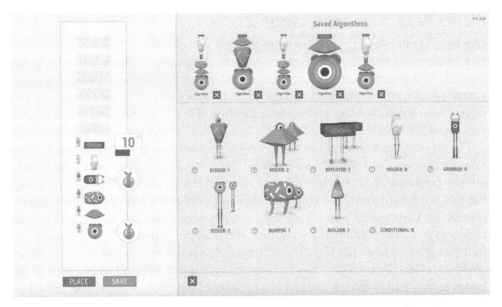

Figure 12.4 The algorithm design screen, where players can access stored algorithms, amend them, rename them, and create new ones.

holes and trenches to reroute water, building up ridges, and finding and collecting resources. Changes to the ecosystem in turn reveal new types of creatures and present new challenges requiring more sophisticated algorithms.

Players can create new algorithms, save and name algorithm configurations, and load them back up to tinker with them in the algorithm design screen (see Figure 12.4). Players can use only those creatures who are already in the Pack when they design algorithms, so they are motivated to engage with, explore, and recruit new creatures. The creatures include the following:

- Diggers, who can dig down one terrain cube per turn
- Movers, who can walk forward one cube per turn
- Repeaters, who can tell the creatures beneath them to repeat their actions a specific number of times
- Holders, who can store resources like fruit
- Grabbers, who can pick up resources like fruit
- Seekers, who can look for a player-specified element in the ecosystem
- Bumpers, who can knock things over
- Builders, who can raise up one terrain cube per turn
- Conditionals (if/then choosers)

An algorithm in this iteration is an ordered stack of creatures that resembles a totem pole, wherein each creature executes its function in sequence. Players can choose where to place the algorithms in the world and when to activate them. Once they are activated the algorithms execute actions dictated by their composition and structure.

Player Responses to Iteration 2

Our second study investigated whether the new first-person, embodied game structure would encourage players' engagement with the game's CT features and whether
the specific interactions of the computational elements with the world's system would
encourage players to reflect on and revise their algorithms. If players acted in these
ways, it would demonstrate a joint understanding of the ecosystem and the functional
aspects of the algorithms.

The players' reactions to Iteration 2 were highly positive and showed much more
extended engagement and exploratory play than in Iteration 1. Some participants were
initially confused about the functions of some of the creatures (such as bumpers),
but this just meant that those creatures were not used much after some initial experimentation. Participants who played long enough were often able to recognize when
creatures became situationally relevant – for example, when a player encountered
fruit that grew on trees too high to pick. Some players did not care what the creatures
could do, at least not initially; they just wanted to find and befriend them all. To do so,
however, they had to master the use of a number of the creatures, and they ended up
learning creature functions along the way.

The fact that the game mixed the design and execution of algorithms did not seem
to overly confuse players. The initial tutorial showed players how to move between
world interfaces and algorithm interfaces. That said, after players left the algorithm design screen, they sometimes forgot that it mattered where they placed the
algorithms in the world, resulting in outcomes like misplaced holes. This is actually
an interesting mistake from a CT perspective, because it helps to illustrate that even a
properly designed algorithm can misbehave if given the wrong input.

Participants understood how to build algorithms, although they needed to experiment to determine how to stack them to attain the desired effect. Younger and older
players responded differently to building and testing algorithms if constraints were
removed. In sessions where players were given immediate access to all pack creatures
rather than having to acquire them gradually through exploration, younger participants
(ages 9–11) tended to produce "kitchen sink" algorithms that threw together all available creatures, whereas older participants (ages 12 and above) tended to differentiate, prune, and whittle down their algorithms to what was necessary for the current
scenario.

Because the game provides a safe, interesting environment to explore, players felt
free to try out different algorithms, "debugging" in an organic fashion when an algorithm did not behave as the player expected. Developing a mastery of all creatures can
take a while for some players, in that many participants were willing to just compen-
sate for their algorithms' shortcomings rather than perfect them. However, this may
be in keeping with CT-STEM, as discussed further below.

The ways in which the process of computation was embodied in Iteration 2 of the
game were successful from an engagement and motivational standpoint, but there
were some unintended consequences of embodying functions in the form of stacks of
creatures. Most 12-year-old players seemed to initially conflate the creature's position

on the stack with its function (top for looking, bottom for walking), based on their existing schema for body-based activity. It took a long period of trial and error for players to change their assumptions about the algorithm stack. One participant said, "Wait a minute. Do I have to put the grabber on the bottom? . . . Because it can't reach [the fruit] all the way from the top." When creating a trench-digging algorithm, another participant struggled with whether to put the digger or the mover on the bottom, exclaiming, "Why did I do that? I forgot to put the mover on the bottom so they move!" When the algorithm moved forward one cube before digging, the player was initially confused but decided it was close enough to his intention, and he kept the design.

Reflections on Integrating Computational and Complex Systems Thinking

Not many educational experiences have tried to explore how CT and complex systems thinking intersect (Berland & Wilensky, 2015). Different methods of engaging learners can have different consequences for the perspectives they develop about computation and complex systems (Berland & Wilensky, 2015), as our two iterations show.

Iteration 1 sought to embody complex system principles in the service of encouraging CT, but it suffered from a number of problems. We wanted our complex system to be manipulable but complex enough to inspire players to use CT to engage and playfully explore a complex environmental system. The elements, while manipulable, did not motivate engagement, as their interactions had no lasting consequences for the environment or the player. Without consequences, there was no natural driver for learners to try to organize their experimentation more efficiently or effectively using CT.

Iteration 2 reversed the strategy and instead embodied computational processes in the service of engaging learners with a complex system. A light narrative framing helped elevate the core goal of exploration, and tying exploration to opportunities to increase one's abilities made it an intrinsically rewarding process.

Conceptually, constructing the stacked Pack algorithms is very similar to blocks-based programming experiences (Wyeth, 2008, Horn et al., 2009), except that, after the algorithm is constructed, the player can place it in the world and release it. Metaphorically, this is similar to instantiating a bit of written code. A similar strategy can be found in agent-based complex systems simulations used in education, such as NetLogo (Resnick & Wilensky, 1998). In these types of simulations, learners program agents and then witness their behavior within the simulated system. Because agents do what they are programmed to do, an agent-based approach brings its own challenges for learner engagement. When learners are deprived of the direct control that is common to virtually all computer games, they can lose interest. To address this, we made sure that the player is another entity in the shared world and can benefit (or suffer) from the changes the algorithms can wreak on the ecosystem. Nothing truly negative or irreversible can happen in the world, but the player can lose time if an algorithm goes astray, much like real-world programmers.

Conclusion

Players explore the world, build their Pack, and, with the help of their Pack, face challenges. Through exploration, experimentation, and iteration, players engage in CT skills and grapple with environmental systems concepts. The design of The Pack evolved into an extended metaphor for the ways in which CT-STEM practitioners approach their work. Typically, CT-STEM professionals develop competency by accident, when they start out wanting to achieve a simple task for which they need to learn some basic computational skills, such as learning how to automate the analysis of temperature sensor data. Once these skills are acquired, professionals might manually reuse their basic analysis program for other sensors before realizing that there might be a way to batch the execution. This necessitates learning more skills. As their computational skill set expands, they see new possibilities for programmatic intervention, and so on in a reciprocal fashion. The Pack makes this reciprocal exploration and accompanying skill set expansion a concrete part of the game.

This is well aligned with Peter Denning's recommendations for encouraging CT: "Adopting computational thinking will happen . . . from making educational offers that help people learn to be more effective in their own domains through computation" (Denning, 2017). A critic of educational programs that attempt to teach CT for the sake of mastering CT, Denning instead advocates that educators "focus on helping students learn to design useful and reliable computations in various domains of interest to them" (Denning, 2017).

Our studies have revealed, however, that in some circumstances, encouraging CT-STEM thinking may run counter to traditional conceptions of CT. The Pack supports habits of mind that are commonly found in CT-STEM but may not fit traditional CT habits of mind. For example, we found that not all players of The Pack immediately seek to develop more efficient algorithms; some make do with "good enough" algorithms that demand more manual engagement. While this might seem like a failure of CT, it may be beneficial for CT-STEM workers, such as scientists who engage in data analytics programming, not to overly automate a process because this provides a closer view of the "terrain" of the work.

There are also benefits to both Pack players and CT-STEM workers in selecting an existing algorithm and tweaking it to adapt to a new challenge rather than going to the trouble of building one from scratch. The library of saved algorithms tangibly represents the player's "toolkit," much like CT-STEM workers maintain personal repositories of useful code. Becoming acquainted with the utility of these CT-STEM practices can help prepare learners for future careers that employ CT in the service of STEM, even though the practices run counter to traditional CT. This may be enough if developing CT-STEM competencies is the goal. However, it suggests that learning environments, curricula, and educators that seek to engage learners in CT through the gateway of CT-STEM should be prepared to highlight when, and why, it would be appropriate to approach problems by more explicitly using traditional CT ideas like abstraction, decomposition, and so on. While not all learners will benefit from mastering CT in the way that computing professionals would (Guzdial, 2015), making learners aware of the advantages of CT is one way to open the door to students who might not otherwise have shown interest in the topic.

leilah lyons et al.

Acknowledgments

The Pack (Integrating Computational Thinking and Environmental Science: Design Based Research on Using Simulated Ecosystems to Improve Student Understanding of Complex System Behavior) was supported through the National Science Foundation (Award No. 1543144). Any opinions, findings, and conclusions or recommendations expressed in this material are those of the authors and do not necessarily reflect the views of the National Science Foundation. The work was further supported by the JPB Foundation. We would also like to thank the following NYSCI researchers who worked tirelessly to bring this project to fruition: Cornelia Brunner, Geralyn Abinader, Amanda Jaksha, and Dorothy Bennett.

References

Ardito, G., Czerkawski, B., & Scollins, L. (2020). Learning computational thinking together: Effects of gender differences in collaborative middle school robotics program. *TechTrends, 64*(3), 373–387. https://doi.org/10.1007/s11528-019-00461-8

Bell, J., & Bell, T. (2018). Integrating computational thinking with a music education context. *Informatics in Education, 17*(2), 151–166.

Berland, M., & Wilensky, U. (2015). Comparing virtual and physical robotics environments for supporting complex systems and computational thinking. *Journal of Science Education and Technology 24*(5), 628–647.

Burke, Q., & Kafai, Y. B. (2012, February). The writers' workshop for youth programmers: Digital storytelling with Scratch in middle school classrooms. In L. S. King & D. R. Musicant (Chairs), *Proceedings of the 43rd ACM Technical Symposium on Computer Science Education* (pp. 433–438). Association for Computing Machinery (ACM). https://dl.acm.org/doi/proceedings/10.1145/2157136

Cuny, J., Snyder, L., & Wing, J. M. (2010). *Demystifying computational thinking for non-computer scientists* [Unpublished manuscript]. www.cs.cmu.edu/~CompThink/resources/TheLinkWing.pdf

Denning, P. J. (2017). Remaining trouble spots with computational thinking. *Communications of the ACM, 60*(6), 33–39. https://doi.org/10.1145/2998438

Grotzer, T. (2012). *Learning causality in a complex world: Understanding of consequence.* Rowman & Littlefield.

Grover, S., & Pea, R. (2013). Computational thinking in K–12: A review of the state of the field. *Educational researcher, 42*(1), 38–43.

Guckenheimer, J., & Ottino, J. M. (2008, September). Foundations for complex systems research in the physical sciences and engineering. Report from an NSF workshop. http://pi.math.cornell.edu/~gucken/PDF/nsf_complex_systems.pdf

Guzdial, M. (2015). Learner-centered design of computing education: Research on computing for everyone. *Synthesis Lectures on Human-Centered Informatics, 8*(6), 1–165.

Horn, M. S., Solovey, E. T., Crouser, R. J., & Jacob, R. J. K. (2009). Comparing the use of tangible and graphical programming languages for informal science education. In D. R. Olsen & R. B. Arthur (Chairs), *Proceedings of ACM CHI 2009 Conference on*

Human Factors in Computing Systems (pp. 975–984). ACM. https://dl.acm.org/doi/10.1145/1518701.1518851

Kafai, Y. B., Lee, E., Searle, K., Fields, D., Kaplan, E., & Lui, D. (2014). A crafts-oriented approach to computing in high school: Introducing computational concepts, practices, and perspectives with electronic textiles. *ACM Transactions on Computing Education (TOCE)*, 14(1), 1–20.

Kelleher, C., Pausch, R., & Kiesler, S. (2007). Storytelling Alice motivates middle school girls to learn computer programming. In M. B. Rossen (Chair), *Proceedings of ACM CHI Conference on Human Factors in Computing Systems* (pp. 1455–1464). ACM. https://doi.acm.org/10.1145/1240624.1240844

Klopfer, E., Yoon, S., & Um, T. (2005). Teaching complex dynamic systems to young students with StarLogo. *Journal of Computers in Mathematics and Science Teaching*, 24(2), 157–178.

Kynigos, C., & Latsi, M. (2007). Turtle's navigation and manipulation of geometrical figures constructed by variable processes in a 3D simulated space. *Informatics in Education International Journal*, 6(2), 359–372.

Maloney, J. H., Peppler, K., Kafai, Y., Resnick, M., & Rusk, N. (2008). Programming by choice: Urban youth learning programming with Scratch. *ACM SIGCSE Bulletin, 1*, 367–371.

Moher, T. (2009). Putting interference to work in the design of a whole-class learning activity. In P. Paolini (Chair), *Proceedings of the 8th International Conference on Interaction Design and Children* (pp. 115–122). ACM. https://dl.acm.org/doi/10.1145/1551788.1551808

National Research Council. (2011). *Report of a workshop on the pedagogical aspects of computational thinking*. The National Academies Press. https://doi.org/10.17226/13170

Papert, S. (1980). *Mindstorms: Children, computers, and powerful ideas*. Basic Books.

Pinkard, N., Erete, S., Martin, C. K., & McKinney de Royston, M. (2017). Digital youth divas: Exploring narrative-driven curriculum to spark middle school girls' interest in computational activities. *Journal of the Learning Sciences*, 26(3), 477–516. https://doi.org/10.1080/10508406.2017.1307199

Repenning, A., Webb, D., & Ioannidou, A. (2010, March). Scalable game design and the development of a checklist for getting computational thinking into public schools. In G. Lewandowsky & S. Wolfman (Chairs), *Proceedings of the 41st ACM Technical Symposium on Computer Science Education* (pp. 265–269). https://doi.org/10.1145/1734263.1734357

Resnick, M. (1994). *Turtles, termites, and traffic jams*. MIT Press.

Resnick, M., & Silverman, B. (2005). Some reflections on designing construction kits for kids. In M. Eisenberg & A. Eisenberg (Chairs), *Proceedings of Interaction Design and Children Conference* (pp. 117–122). https://doi.org/10.1145/1109540.1109556

Resnick, M., & Wilensky, U. (1998). Diving into complexity: Developing probabilistic decentralized thinking through role-playing activities. *Journal of the Learning Sciences 7*, 153–172.

Salac, J., & Franklin, D. (2019). *Why access isn't enough: An analysis of elementary-age students' computational thinking performance through an equity lens.* University of Chicago. https://newtraell.cs.uchicago.edu/files/ms_paper/salac.pdf

Scott, A., Martin, A., McAlear, F., & Madkins, T. C. (2016). Broadening participation in computer science: Existing out-of-school initiatives and a case study. *ACM Inroads, 7*(4), 84–90.

Selby, C., & Woollard, J. (2014). *Refining an understanding of computational thinking.* University of Southampton Institutional Repository. https://eprints.soton.ac.uk/372410/

Shelley, T., Lyons, L., Zellner, M. L., & Minor, E. (2011). Evaluating the embodiment benefits of a paper-based TUI for educational simulations. In *CHI '11: extended abstracts on Human Factors in Computing Systems* (pp. 1375–1380). ACM. https://doi.org/10.1145/1979742.1979777

Tisue, S., & Wilensky, U. (2004, October). NetLogo: Design and implementation of a multi-agent modeling environment [Paper presentation]. Agent 2004, Chicago, IL. http://ccl.northwestern.edu/papers/2013/netlogo-agent2004c.pdf

Torres-Torres, Y. D., Román-González, M., & Pérez-González, J. C. (2020). Unplugged teaching activities to promote computational thinking skills in primary and adults from a gender perspective. *IEEE Revista Iberoamericana de Tecnologias del Aprendizaje, 15*(3), 225–232.

Weintrop, D., Beheshti, E., Horn, M., Orton, K., Jona, K., Trouille, L., & Wilensky, U. (2016). Defining computational thinking for mathematics and science classrooms. *Journal of Science Education and Technology, 25*(1), 127–147.

Wing, J. (2006). Computational thinking. *Communications of the ACM, 49*(3), 33–35.

Wing, J. (2011, Spring) Research notebook: Computational thinking – what and why? *The Link, 6,* 20–23. Carnegie Mellon University. http://link.cs.cmu.edu/files/11-399_The_Link_Newsletter-3.pdf

Wyeth, P. (2008). How young children learn to program with sensor, action, and logic blocks. *Journal of the Learning Sciences 17*(4), 517–550.

Yoon, S. A., Anderson, E., Koehler-Yom, J., Evans, C., Park, M., Sheldon, J., . . . Klopfer, E. (2017). Teaching about complex systems is no simple matter: Building effective professional development for computer-supported complex systems instruction. *Instructional Science, 45*(1), 99–121.

Yoon, S. A., Goh, S. E., & Park, M. (2018). Teaching and learning about complex systems in K–12 science education: A review of empirical studies 1995–2015. *Review of Educational Research, 88*(2), 285–325.

Index

Note: Page references for figures are in *italics* and references to tables are in **bold**.

agency: defined 62–3; within the design process 62–3; engagement and 181; during engineering activity development and research 56; family groups in maker spaces 129–32; of Explainers 31–2, 34, 37–9; and identity development 18–20; in informal STEM learning 14; learner agency 23; limitations on 15–16; for non-native English speakers 161–9; participatory educational approaches to 16–18; relational agency 62–3; relational agency in the Pokédex Case 69–75; relational research-practice partnerships 75–7; situated-relational design 63, 65–9; through embodied learning 174; visitor agency for family groups 20, 21–3, 87–8

Big Data for Little Kids (BDLK) program: caregiver engagement in 22, 107–16; caregivers in informal STEM settings 22, *22*, 101–2; community engagement in 107–9; data modeling in informal settings 102–3, 106–7, **107**, *108*; Design Make Play within 103; goals of 100–1; implementation 105–6; program development 103; prototype activities and workshops 103–5, **104**, *105*; research background 101–3

caregivers: engagement and agency of in museum settings 21–3, 87–8; engagement in the Big Data for Little Kids program 22, 107–16; goals in museum settings 21; interactions with facilitators 21–2; learning and cultural practices 13–14, 21; non-English speaking caregivers 88, 94, 95, 113; role in informal STEM experiences 22, *22*, 101–2, 106, *108*; *see also* family groups
Center for Children and Technology (CCT) 6–7
Coalition for a STEM Future 96–7
communities: Corona, Queens demographic 83–4, 86, 121, 155–6, 158, 215; design-thinking for problem-solving 136; museum-community engagement initiatives 84, 87–9; role of the local community for the Innovation Institute project 139–40, 145–6; youth social entrepreneurship 135–6; *see also* museum-community engagement; NYSCI Neighbors Initiative
computational thinking skills: and complex systems simulations 212–13, 221;

definitions 196, 197–9, 210–11; designs
to instill computational thinking 200–2;
example of 195–6; implementation review
for elementary schools program 204–5;
Integrating Computational Thinking initiative
197; learner engagement with 211; low-
income students access to 208–9; in the
maker movement 136–7; NYSCI's coaching
model in computational thinking 198–9; in
The Pack 208; for problem-solving in the
Innovation Institute project 136–7, 144–5,
145; questioning strategies for 201–2, **201**; in
STEM education 196–7, 209–10, 211–12, 222;
use by English Language Learners 203–4;
use of decimals case study 202–3
computer science education 196
Connected Worlds: New York Hall Of Science
(NYSCI) 30, 61, 63, *64*; the Pokédex Case
69–75; role of the Explainers 63, 64, 65; *see
also* digital dashboard; The Pack
cultural values: engagement of culturally
diverse family groups 23, 84, 85–6, 92,
107–16; learning and cultural practices 13–14,
21; multicultural programming 93–4; STEM
disciplines as cultural practices 4, 7, 21, 23,
101

Design Lab: background to 30, 31; Design Make
Play within 32; engineering activities 31, 33,
38, 40, 46–7, 56–7; *see also* Explainers
Design Make Play: in the Big Data for Little
Kids (BDLK) program 103; core principles
3; in the Design Lab 32; ethos of 1–4;
Explainers' role 31–2, 39; at the NYSCI 1–2,
29–30; in the NYSCI Neighbors Initiative 87;
in Playground Physics 177–8
design processes: for the Big Data for Little
Kids (BDLK) program 101–5, 111–16;
brainstorming and eliciting tasks 70, *70*;
co-design strategies for the digital dashboard
63–9; collaboration for staff agency 56;
collaborative research and development
processes 48–51, 55–6, 64, 65; data analytics
and prototypes for the digital dashboard
72–4, *73, 74*; within the Design Make
Play ethos 2; of the digital dashboard for
Connected Worlds exhibit 61–2, 63–4; Digital
Design for English Language Learners
157–9, 160–1; early testing and prototypes
for the digital dashboard 74–5; engineering
design activities 19, 40; engineering in the
Design Lab 31, 33, 38, 40, 46–7, 56–7; of the
Innovation Institute (I2) program 138–40,

138, 144; Maker Space programs 123–5,
127–9; museum-community engagement
initiatives 84, 88–9; narratives for empathic
design practices 46–8, 51–2, 53–5, 56–7;
NYSCI's coaching model in computational
thinking 198–9; The Pack 213–21; the
Pokédex Case 69–75; relational research-
practice partnerships 75–7; relationship
and agency in 62–3; scenarios and design
sketches 71–2; situated-relational design 63,
65–9
digital dashboard: as a collaborative co-design
project 63–9; for the Connected Worlds
exhibit 61–2, 63–4; Explainers' involvement
in the research process 65–6, 70–5, 76–7;
Explainers' use of 64; focus study groups
66; interviews 66; observation groups 65–6;
participatory design sessions 66–8, **67–8**, *69*
Digital Design for English Language Learners
(ELLs): agency and inclusion benefits for
ELL 161–7; background to 155–6; challenges
of STEM learning for ELL 156; ELL students
159, 161–7; ELL teachers 157–9, 169–70;
embodied learning strategies 156; Explainers'
role in 159–60; interdisciplinary team
158–60; multimodal approaches 160–1;
Noticing Tools 161, 162, *162, 163, 164, 165*;
parental involvement 170; potential of digital
tools for ELL STEM learning 156–8; teacher
observers 159, 171
digital resources: Choreo Graph 156, 157, 162,
162, 163, 165; computational thinking skills
and 198; EcoMuve 213; engagement in
science learning 176; Fraction Mash 156, 157,
164; in the Innovation Institute project
143–8, **145**; in The Pack 208; in the
Playground Physics project 176, 178–9,
178, 182–3, 186; use by English Language
Learners 157; *see also* The Pack

embodied learning strategies: applied to STEM
learning 174; in digital games 213; EcoMuve
213; embodied play and playfulness 175–6,
180; for English Language Learners 156; in
informal settings 173–4; in The Pack 209–10,
220–1; Science Playground exhibit 176–7, *177*
engagement: agency of caregivers in museum
settings 21–3, 87–8; caregiver engagement
with the Big Data project 22, 107–16; with
computational thinking 211; of culturally
diverse family groups 23, 84, 85–6, 92,
107–16; family engagement in maker
spaces 122–3, 125–7; female engagement in

engineering 19, 45, 48, 53; in formal settings 180–1; in informal settings 181; museum-community engagement initiatives 84, 88–9; in the Playground Physics project 184–8; productive engagement strategies 186–8; *see also* NYSCI Neighbors Initiative

engineering: activity development and research 48–51, 55–6; in the Design Lab 31, 33, 38, 40, 46–7, 56–7; empathy as integral to 19, 44–6; engineering design activities 19, 40; Explainers' roles in narrative framings 49, 55, 56; female engagement in the field of 19, 45, 48, 53; habits of mind strategies for 33, 38; Help Grandma activity 44, 52, 53, 54; Help the Pets activity 52, 55; narrative/activity pairs 51–2; narratives for empathic design practices 46–8, 51–2, 53–5, 56–7; visitor engagement with engineering practices 32

English Language Learners (ELL): challenges of STEM learning 156; computational thinking skills for 203–4; embodied learning strategies 156; potential of digital tools for 157; teachers in informal learning settings 170; *see also* Digital Design for English Language Learners

Explainers: agency for 31–2, 34, 37–9; apron tools 40; changing role of 29, 30, 32–3, 37–8, 41–2, 76; in the Connected Worlds exhibit 63, 64, 65; content knowledge 39–40; creation of visitor agency 30; within the Design Make Play ethos 31–2, 39; in the Digital Design for English Language Learners 159–60; facilitation frameworks 40; habits of mind strategies for 33–6, 39, 40, 42; interdepartmental collaboration with 37, 38–9; narrative framing for engineering activities 49, 55, 56; orientation for new Explainers 32–3, 34, 36; within the research and development of the digital dashboard 65–6, 70–5, 76–7; Science Career Ladder (SCL) 30–1; science demonstrations 38; talk-back boards 34, *36*, 39; traditional training program 28, 31

facilitators: Innovation Institute (I2) 142; interactions with caregivers 21–2; Maker Space programs 17, 123, 125, 126, 128–9; *see also* Explainers

family groups: within community engagement strategies 87–8; engagement of culturally diverse family groups 23, 84, 85–6, 92, 107–16; engagement with maker spaces 122–3, 125–7; family group agency in maker spaces 129–32; family STEM learning 90–1, 91; fostering engagement in the Big Data project 107–16; institutional trust 88, 92–3, 110; intergenerational learning spaces 90, 92, 95; parent empowerment 92–3; parent workshops 91–2; visitor agency for 20, 21–3; *see also* NYSCI Neighbors Initiative

Innovation Institute (I2): collaborative engagement work 139, *140*; community engagement and problem-solving 140, 145–6; computational making and problem-solving 136–7, 144–5, **145**; design-thinking for problem-solving 136; as an emergent learning environment 141; exploration of the community 139–40; facilitators 142; goals of 20, 135; problem-solving practices 146–9; program design 138–40, *138*, 144; reflection booth 143; research on youth engagement with 141; research–practice partnership 142; social entrepreneurship and youth development 135–6; theoretical framework 135–7, *137*; 2019 implementation case study 143–6

Integrating Computational Thinking initiative 197

knowledge transmission: participatory educational approaches to 16–18; traditional approaches to 14, 16

language: embodied learning strategies 156; English as a Second Language (ESL) classes 88; language accessibility at NYSCI 93, 113; multilingual Explainers 159–60; non-English speakers in the NYSCI community 83, 85, 86, 94; non-English speaking caregivers 88, 94, 95, 113; school transition for English Language Learners 92; *see also* Digital Design for English Language Learners; English Language Learners

learning experiences: identity development through 18–19, 20; importance of learner agency to 14–15; in informal STEM settings 13–14; narratives as mechanisms for 46–8

maker movement: computational making and problem-solving 136–7; emergence of 29; participatory practices in 17

Maker Space programs: access to new tools 128; case studies on family group agency as learners 129–32; collaborative art pieces 124; design process 123–5, 127–9; distribution of materials 128; facilitators 17, 123, 125, 126, 128–9; family engagement with 122–3, 125–7; implementation 126; Marble Run 123; pedagogical approach 121–2; prototype testing 125; room set up 127–8; for STEM learning 120–1; tabletop activities 123, 125, 126; tool-based walk-up activities 123–5, *124*

mathematics: Choreo Graph (app) 156, 157; data education 102; data modeling in informal settings 102–3, 106–7, **107**, *108*; digital tools for ELL STEM learning 157–8; Fraction Mash (app) 156, 157; *see also* Big Data for Little Kids (BDLK) program; Digital Design for English Language Learners

museum-community engagement: as alternative sites for STEM engagement 84–5; for diverse communities 85–6; *see also* Big Data for Little Kids (BDLK) program; communities; NYSCI Neighbors Initiative

New York Hall Of Science (NYSCI): Connected Worlds 30, 61, 63, *64*; Corona, Queens demographic 83–4, 86, 121, 155–6, 158, 215; Design Make Play at 1–2, 29–30; school collaboration 197; Science Playground exhibit 176–7, *177*; Small Discoveries exhibit 18; visitor-centered STEM engagement 28–9; *see also* NYSCI Neighbors Initiative

Noticing Tools: Choreo Graph 156, 157, 162, *162*, *163*, *165*; Fraction Mash 156, 157, 164; use by English Language Learners 157

NYSCI Neighbors Initiative: challenges 96; community communication initiatives 88–9; community engagement activities 90–4; Corona, Queens demographic 83–4, 86; Design Make Play within 87; ecologies of care framework 86; engagement and agency of family groups 87–8; family STEM learning 90–1, *91*; free after-school programs 90; future directions 95–6; impact of community engagement activities 94–5; institutional trust building 88, 92–3, 110; intergenerational learning spaces 90, 92, 95; language accessibility 93; multicultural programming 93–4; within the national setting 96–7; outreach programs 89–90; parent workshops 91–2; STEM engagement initiatives 84; strategies for community engagement 87–90

The Pack: algorithm design functionality 218–19, *219*, 220, 221; background to 209–10, *209*, 212; computational thinking

skills and 208; design partners 214; design process 213–21; elemental avatars 216, *216*, *217*; embodied game structure 220–1; iterative development process 214–15; player engagement with 220–1; prototype testing 215; testing and feedback on the first iteration 216–18, *217*, 221; testing and feedback on the second iteration 218–20, *219*; use of embodied learning strategies 209–10, 220–1; users complex causality exploration 216–18

Playground Physics: background to 174–5; collaborative research and development process 179–80; curriculum 179; Design Make Play within 177–8; embodied play and playfulness 175–6, 180; evidence of engagement 184–6; implementation challenges 184–7; implementation in a school 180, 181–3; observation protocols 181; productive engagement strategies 186–8; teacher support activities 179; use of digital tools 176, 178–9, *178*, 182–3, 186

research: for the Big Data for Little Kids (BDLK) program 101–5, 107–11; brainstorming and eliciting tasks 70, *70*; collaborative research and development processes 48–51, 55–6, 64, 65; collaborative research in the Playground Physics program 179–80; data analytics and prototypes 72–4, *73*, *74*; data modeling in informal settings 103–4; early testing and prototypes 74–5; Explainers' involvement in the digital dashboard research process 65–6, 70–5, 76–7; on family engagement with maker spaces 122–3; focus study groups 65, 66; on the Innovation Institute (I2) 141–3; interviews 66; museum-community engagement initiatives 88–9; observation groups 65–6; participatory design sessions 66; the Pokédex Case 69–75; potential of tools for ELL STEM learning 157; relational research-practice partnerships 75–7; scenarios and design sketches 71–2; within situated-relational design 63, 65–9

science education: empowering STEM identities 20; identity development through 19–20; participatory educational approaches to 16–18; role of learner agency in 14–15; traditional approaches to knowledge transmission 14, 16

social entrepreneurship 135–6

STEM (science, technology, engineering, and mathematics): challenges for English Language Learners 156; computational thinking skills and 196–7, 209–10, 211–12, 222; computer science courses 196; Design Make Play within 1–4; digital tools for ELL STEM learning 156–8; embodied learning strategies and 174; engagement challenges in formal learning environments 174; family STEM learning 90–1, 91; learning experiences in informal STEM settings 13–14; STEM disciplines as cultural practices 4, 7, 21, 23, 101; underrepresented groups in 134–5; *see also* engineering; mathematics